Nanogrids, Microgrids, and the Internet of Things (IoT)

Nanogrids, Microgrids, and the Internet of Things (IoT): Towards the Digital Energy Network

Special Issue Editor

Antonio Moreno-Munoz

MDPI • Basel • Beijing • Wuhan • Barcelona • Belgrade

MDPI

Special Issue Editor
Antonio Moreno-Munoz
University of Cordoba
Spain

Editorial Office
MDPI
St. Alban-Anlage 66
4052 Basel, Switzerland

This is a reprint of articles from the Special Issue published online in the open access journal *Energies* (ISSN 1996-1073) from 2018 to 2019 (available at: https://www.mdpi.com/journal/energies/special_issues/nanogrids)

For citation purposes, cite each article independently as indicated on the article page online and as indicated below:

LastName, A.A.; LastName, B.B.; LastName, C.C. Article Title. *Journal Name* **Year**, *Article Number*, Page Range.

ISBN 978-3-03921-794-6 (Pbk)
ISBN 978-3-03921-795-3 (PDF)

Contents

About the Special Issue Editor

Antonio Moreno-Munoz is a Professor at the Department of Electronics and Computer Engineering, Universidad de Córdoba, Spain, where he is the Chair of the Industrial Electronics and Instrumentation R&D Group. He received his Ph.D. and M.Sc. degrees from UNED, Spain in 1998 and 1992, respectively. From 1981 to 1992, he was with RENFE, the Spanish National Railways Company. Since 1992, he has been with the University of Cordoba, where he has been the director of its department and academic director of the Master in Distributed Renewable Energies.

His research focuses on Smart Cities, smart grids, power quality, and the Internet of Energy. He has participated in 20 projects and/or R&D contracts, including the management of 9 of them, and he has more than 130 publications in these topics.

He is a Member of European Technology & Innovation Platforms (ETIP) Smart Networks for Energy Transition (SNET) WG4, a senior member of the "Technical Committee on Smart Grids" IEEE IES, and a member of the CIGRÉ/CIRED JWG-C4.24 committee "Power Quality and EMC Issues associated with future electricity networks". He has also been a member of the IEC/CENELEC TC-77/SC-77A/WG-9 committee, as well as a member of the ISO International Organization for Standardization AEN/CTN-208/SC-77-210.

He is an evaluator of R&D&I projects for the Estonian Research Council, the Fund for Scientific and Technological Research (FONCYT) of the National Agency for the Promotion of Science and Technology in Argentina and the Directorate General of Research, Development, and Innovation of the Ministry of Science, Innovation, and Universities of Spain. He is also an evaluator for European Quality Assurance (EQA) and DNV-GL.

Other roles include: Springer Science editorial consultant. Associate editor in the *Electronics* and *Energies* journals of MDPI-AG. Editor in the *Intelligent Industrial Systems* journal, (Springer Science), *Journal of Energy*, Hindawi, Editor in *The Scientific World Journal* Hindawi, Review Editor, *Frontiers in Energy Research*, and *Energy Systems and Policy*. In addition to this, he has served as a guest editor and a regular reviewer for several IEEE, IET, MDPI, and Elsevier journals.

Preface to "Nanogrids, Microgrids, and the Internet of Things (IoT): Towards the Digital Energy Network"

In the coming years, ICTs aim to converge with energy systems to make them smarter, more efficient, and more reliable. With the advent of embedded submetering, the energy information available will become more granular and more easily attributable to specific appliances or systems. The proliferation of thousands of IoT sensors throughout the power grid will create volumes of information about usage patterns, failures or the state of assets hitherto inconceivable. Rapid advances in machine learning and big data analytics are enabling the emergence of a wide range of disruptive energy products and personalized services that will benefit and empower digital citizens. This whole transition from the smart grid toward the digital energy platform is also posing new challenges to security and privacy. Definitively, this will change markets, business, and employment. Let us turn the page and dive into this universe; the adventure has only just begun.

Antonio Moreno-Munoz
Special Issue Editor

energies

MDPI

Editorial

Special Issue "Nanogrids, Microgrids, and the Internet of Things (IoT): Towards the Digital Energy Network"

Antonio Moreno-Munoz

Department of Electronics and Computer Engineering, University of Cordoba, E-14071 Cordoba, Spain; el1momua@uco.es

Received: 16 September 2019; Accepted: 10 October 2019; Published: 14 October 2019

1. Introduction

I started hearing a lot about energy digitization a little over a year ago, talking to my colleagues during conferences and meetings. People did not agree with what they meant by "digitization". In my opinion, the power grid has long since ceased to be analog. The smart grid has always essentially been digital. So why this term, and why now? Is it just one of those buzzwords that are used for marketing purposes and does not really mean much at all? Once the word "smart" was exhausted, the technological gurus had to find another one that groups enough novelties and was broad as well. May not has been the right choice, but here we are, now with the need to give content or highlight some nuances that were no longer present in the pioneering smart grid concept.

A good approximation is the one we have used in the European Technology and Innovation Platform for Smart Networks for the Energy Transition Working Group on Digital Energy (ETIP SNET WG4). According to our position paper [1], digitization consists of "using digital technologies in order to change business models and provide new revenue streams and value-producing opportunities". This means that the Internet of Things, machine (M2M) communications, artificial intelligence, machine learning, digital twin, and other developments must find their use in the energy system. Indeed, the digitization of the energy system is not a recent development. Driven by new regulations, new market structures, and new energy resources, the smart grid has in recent decades been the trigger for profound changes in the way electricity is generated, distributed, managed and consumed. The smart grid has raised the traditional power grid by using two-way electricity and information flows to create an advanced, automated power supply network. Seen with insight, it could be said that smart grids have been the flagship application of digital technology in the energy sector. However, so far, the focus has been on the operation of the infrastructure. In the future energy systems, smart meters and appliances will expand the demand-response potential, enabling new relationships between the end-user and the energy system, and where service will be key.

In today's technological landscape, we can access feasible data and knowledge that has so far been merely inconceivable. This special issue aims to address this landscape towards which, the smart grid is progressing due to the advent of all these pervasive technologies. It will be the advanced exploitation of the massive data generated from internet of things (IoT) sensors that will become the main driver in evolving the concept of smart grids towards a digital energy network paradigm, focused on service. Furthermore, collective intelligence will improve the processes of decision making and empower citizens.

This issue includes outstanding cases of the use of these technologies and their advantageous energy system application at different levels, such as generation, distribution, or consumption.

2. A Short Review of the Contributions in This Issue

The paper "*A Novel Direct Load Control Testbed for Smart Appliances*" [2] proposes a novel Direct Load Control (DLC) testbed, aiming to conveniently support the research community, as well as analyzing and comparing their designs in a laboratory environment. Based on the LabVIEW computing platform, this original testbed enables access to knowledge of major components such as online weather forecasting information, distributed energy resources (e.g., energy storage, solar photovoltaic), dynamic electricity tariff from utilities and demand response (DR) providers, together with different mathematical optimization features given by a General Algebraic Modelling System (GAMS). This intercommunication is possible thanks to the different applications' programming interfaces (API) incorporated into the system and to intermediate agents specially developed for this case. Different basic case studies have been presented to envision the possibilities of this system in the future and more complex scenarios, to actively support the DLC strategies. These measures will offer enough flexibility to minimize the impact on user comfort combined with support for multiple DR programs. Thus, given the successful results, this platform can lead towards a solution that has more efficient use of energy in the residential environment.

The paper "*An Alternative Internet-of-Things Solution Based on LoRa for PV Power Plants: Data Monitoring and Management*" [3] proposes a wireless low-cost solution based on long-range (LoRa) technology able to communicate with remote photovoltaic (PV) power plants, covering long distances with minimum power consumption and maintenance. This solution includes a low-cost open-source technology at the sensor layer and a low-power wireless area network (LPWAN) at the communication layer, combining the advantages of long-range coverage and low power demand. Moreover, it offers an extensive monitoring system to exchange data in an Internet-of-Things (IoT) environment. A detailed description of the proposed system at the PV module level of integration is also included in the paper, as well as detailed information regarding LPWAN application to the PV power plant monitoring problem. In order to assess the suitability of the proposed solution, results collected in real PV installations connected to the grid are also included and discussed.

The paper "*Advantages of Minimizing Energy Exchange Instead of Energy Cost in Prosumer Microgrids*" [4] two novel approaches are proposed: Firstly, a different objective function, relative to the mismatch between generated and demanded power is tested and compared to a classical objective function based on energy price, by means of a genetic algorithm method; secondly, this optimization procedure is applied to batteries' coordinated scheduling of all the prosumers composing a community, instead of single one, which better matches the microgrid concept. These approaches are tested on a microgrid with two household prosumers, in the context of Spanish regulation for self-consumption. Results show noticeably better performance of mismatch objective functions and coordinated scheduling, in terms of self-consumption and self-sufficiency rates, power and energy interchanges with the main grid, battery degradation, and even economic benefits.

In the paper "*An Analysis of Voltage Quality in a Nanogrid during Islanded Operation*" [5] voltage quality data has been collected in a single house nanogrid during 48 weeks of islanded operation and 54 weeks of grid-connected operation. The voltage quality data contains the voltage total harmonic distortion (THD), odd harmonics three to 11 and 15, even harmonics four to eight, voltage unbalance, short-term flicker severity (Pst) and long-term flicker severity (Plt) values, and voltage variations at timescales below 10 min. A comparison between islanded and grid-connected operation values was made, where some of the parameters were compared to relevant grid-standard limits. It is shown that some parameters exceed the defined limits in the grid-standards during islanded operation. It was also found that the islanded operation has two modes of operation, one in which higher values of the short circuit impedance, individual harmonic impedance, harmonic voltage distortion, and voltage unbalance were reached.

In the paper "*An Analysis of Frequency Variations and its Implications on Connected Equipment for a Nanogrid during Islanded Operation*" [6] frequency, voltage and reliability data have been collected in a nanogrid for 48 weeks during islanded operation. Frequency values from the 48-week measurements were analyzed and compared to relevant limits. During 19.5% of the 48 weeks, the nanogrid had curtailed the production due to insufficient consumption in islanded operation. The curtailment of production was also the main cause of the frequency variations above the limits. When the microgrid operated on stored battery power, the frequency variations were less than in the Swedish national grid. Of all the interruptions that occurred in the nanogrid, 39.4% are also indirectly caused by the curtailment of solar production

A smart inverter should offer some features such as plug and play, self-awareness, adaptability, autonomy, and cooperativeness. These features are introduced and comprehensively explained in the paper "*Smart Inverters for Microgrid Applications: A Review*" [7]. In order to achieve higher functionality, efficiency and reliability, in addition to improving the control algorithms it is beneficial to equip the inverters with "smart" features. One interpretation of "smartness" refers to minimizing the requirement of communication and therefore switching from centralized to decentralized control. At the same time, being equipped with efficient and state of the art communication protocols also indicates "smartness" since the requirement of communication cannot be completely omitted. One contribution discussed here is the possibility of achieving long-range wireless communication between inverters to empower various control schemes. Although current efforts aim to modify and improve power converters in a way that they can operate communication free, if a suitable and functional communication protocol is available, it will improve the accuracy, speed, and robustness of them.

Conflicts of Interest: The authors declare no conflict of interest.

References

1. ETIP SNET WG4. *Digitalization of the Energy System and Customer Participation.* 2017. Available online: https://www.etip-snet.eu/wp-content/uploads/2018/11/ETIP-SNET-Position-Paper-on-Digitalisation-short-for-web.pdf (accessed on 13 October 2019).
2. Garrido-Zafra, J.; Moreno-Munoz, A.; Gil-de-Castro, A.; Palacios-Garcia, E.J.; Moreno-Moreno, C.D.; Morales-Leal, T. A Novel Direct Load Control Testbed for Smart Appliances. *Energies* **2019**, *12*, 3336. [CrossRef]
3. Paredes-Parra, J.M.; García-Sánchez, A.J.; Mateo-Aroca, A.; Molina-Garcia, A. An Alternative Internet-of-Things Solution Based on LoRa for PV Power Plants: Data Monitoring and Management. *Energies* **2019**, *12*, 881. [CrossRef]
4. González-Romera, E.; Ruiz-Cortés, M.; Milanés-Montero, M.I.; Barrero-González, F.; Romero-Cadaval, E.; Lopes, R.; Martins, J. Advantages of Minimizing Energy Exchange Instead of Energy Cost in Prosumer Microgrids. *Energies* **2019**, *12*, 719. [CrossRef]
5. Nömm, J.; Rönnberg, S.; Bollen, M. An Analysis of Voltage Quality in a Nanogrid during Islanded Operation. *Energies* **2019**, *12*, 614. [CrossRef]
6. Nömm, J.; Rönnberg, S.; Bollen, M. An Analysis of Frequency Variations and its Implications on Connected Equipment for a Nanogrid during Islanded Operation. *Energies* **2018**, *11*, 2456. [CrossRef]
7. Arbab-Zavar, B.; Palacios-Garcia, E.; Vasquez, J.; Guerrero, J. Smart Inverters for Microgrid Applications: A Review. *Energies* **2019**, *12*, 840. [CrossRef]

Review

Smart Inverters for Microgrid Applications: A Review

Babak Arbab-Zavar, Emilio J. Palacios-Garcia *, Juan C. Vasquez and Josep M. Guerrero

Department of Energy Technology, Aalborg University, Pontoppidanstraede 111, DK-9220 Aalborg, Denmark; baz@et.aau.dk (B.A.-Z.); juq@et.aau.dk (J.C V.); joz@et.aau.dk (J.M.G.)
* Correspondence: epg@et.aau.dk; Tel.: +5-5028-0730

Received: 6 February 2019; Accepted: 24 February 2019; Published: 4 March 2019

Abstract: In a microgrid, with several distributed generators (DGs), energy storage units and loads, one of the most important considerations is the control of power converters. These converters implement interfaces between the DGs and the microgrid bus. In order to achieve higher functionality, efficiency and reliability, in addition to improving the control algorithms it is beneficial to equip the inverters with "smart" features. One interpretation of "smartness" refers to minimizing the requirement of communication and therefore switching from centralized to decentralized control. At the same time, being equipped with efficient and state of the art communication protocols also indicates "smartness" since the requirement of communication cannot be completely omitted. A "smart inverter" should offer some features such as plug and play, self-awareness, adaptability, autonomy and cooperativeness. These features are introduced and comprehensively explained in this article. One contribution discussed here is the possibility of achieving long-range wireless communication between inverters to empower various control schemes. Although current efforts aim to modify and improve power converters in a way that they can operate communication free, if a suitable and functional communication protocol is available, it will improve the accuracy, speed and robustness of them.

Keywords: smart inverter; microgrid; distributed generation; communication; wireless

---

1. Introduction

Microgrids are a form of small-scale grids that contain DGs, energy storage units and linear or nonlinear loads that can operate in grid-connected or islanded mode. In microgrids, DGs can be of renewable or non-renewable nature, and the components of such grids are interfaced by power converters [1–3]. The CIGRE working group C6.22 Microgrid Evolution Roadmap (WG6.22) provides a standard definition for microgrids: Microgrids are electricity distribution systems containing loads and distributed energy resources, (such as distributed generators, storage devices, or controllable loads) that can be operated in a controlled, coordinated way either while connected to the main power network or while islanded [4]. In the early years of introducing renewable energy sources (RES), the power generated by them was not substantial in comparison to the large conventional generators powering up the grid, therefore, their impact on the performance of the grid was almost unnoticeable. At those years, the scheme was to allow them to produce as much power as possible and inject it into the grid by using their own algorithms. The main conventional generators could regulate the small unbalances and fluctuations caused by these DGs [5,6].

However, due to the recent development and expansion of renewable energy technologies, RESs have become a major energy source in some countries. Therefore, the quality and parameters of their injected power into the grid must be carefully monitored. In other words, their operation needs to be accompanied by power management schemes centered on power sharing or load sharing to be fully controllable [7].

The concept of power sharing between parallel DGs was initially introduced for synchronous generators in large-scale grids, and recently some novel approaches were proposed to incorporate the idea of droop control algorithms into microgrids [8–13]. Although using droop-based methods can be very beneficial, some drawbacks can be identified. For example, the trade-off between voltage regulation and load sharing [14], the ineffective management of harmonics introduced by nonlinear loads [15] and the slow dynamic response due to the incorporation of low pass filters for calculating average values of active (P) and reactive (Q) power [16]. Because of the problems of conventional droop control, its principles have undergone numerous modifications and improvements. In [17] a new droop scheme aiming to regulate the voltages of each inverter and improve current sharing was proposed. When nonlinear loads such as rectifiers are present, they introduce harmonics in the system, and prevent droop control from working efficiently. In [18], this problem was addressed and regulated by adding harmonic droop characteristics. This issue can also be compensated by using a virtual variable impedance in series with the load. As its name indicates, this impedance is purely simulated and no physical components are added to the circuit [19–22]. In [13] a hierarchical control method was proposed to control microgrids. This hierarchy consists of three layers: the primary level is droop control; the secondary level compensates for the unbalances and deviations resulted by the first level by sending voltage and frequency references; and the tertiary level (energy management level) which sends the droop coefficients is where the connection with the grid. In this method, only the first level works as communication free, since it is droop control, while the rest of them require communication between inverters.

Although most of the existing microgrids are supplied in AC to simplify their interface with utility grids, DC microgrids are attracting more attention in recent years. This is not only because of the DC nature of several RESs such as solar energy, fuel cells and energy storage units like batteries and supercapacitors, but also because of some additional benefits. In a DC microgrid, the problem of harmonic current sharing no longer exists. Neither is there reactive power sharing. What is more, the system can be more efficient and simple as there is no more need for AC-DC-AC converters. These converters are commonly used for wind or hydro energy conversion and they consist of two back-to-back bidirectional AC-DC converters with a DC link. This DC link decouples two 3-phase AC systems with different parameters enabling the maximization of the output power by using maximum power point tracking (MPPT) algorithms and the synchronization with the grid [12,13,23–26]. The hierarchical control for DC microgrids is similar to the one proposed for AC but simpler. At the primary level, the droop control of DC microgrids only consists of voltage versus current droop in contrast to AC microgrids where active and reactive power droop controls must be provided. The secondary control regulates voltage deviations resulted from the primary control and restores the values to nominal levels, which is only applied at islanded mode. In grid-connected mode, the references comes from the grid parameters. Similar to AC microgrids the tertiary control regulates the power flow with a stiff DC grid at energy management level [13,27,28].

The power converters are able to play different roles when incorporated into a microgrid, namely grid forming, grid feeding and grid following. These are defined in [10] alongside with control structures at the converter level. Another fact that should be considered is that DGs in a microgrid are commonly placed far away from each other. Even though using various droop control schemes reduces the requirement of communication between inverters, a link is unavoidable if a good power sharing control scheme is desired. However, to implement fast communication links which can fulfill the requirements of this kind of applications and be sufficiently reliable can be very costly [10,29]. Therefore, finding a long range and preferably low operating cost communication protocol to fill in for the current lack of effective bidirectional communication between inverters can be considered a very promising field for research and development. In this paper, a literature review on "smart inverters" and their application to microgrids is conducted. The "smartness" features are introduced and explained in detail. In each section, different methods and challenges regarding each of these indicators are addressed and are empowered by equations and tabular or illustrative information where required. The technological achievements, as well as present gaps, are introduced throughout the article. Finally, the discussion section presents the main conclusions drawn from the study and provides insights into possible paths for further research and development.

2. Smart Inverters

Defining a device as "smart" means that it has the ability to operate efficiently and autonomously with limited operator intervention required. The role of an inverter in a microgrid is to operate as an interface between energy generation and consumption points. Therefore, its role is not limited to AC DC conversion or vice versa (depending on the type of the converter) but also to control the power flow, sense faults, disconnect when necessary and other functions. Since these are the main factors that concern the microgrid control, it can be stated that the inverters are the thinking and processing components of a microgrid, which collect data and configure themselves in order to operate in a safe, controlled and effective environment. Furthermore, each power converter acting as an interface for a distributed energy resource (DER) to the grid has to fulfill the requirements of the IEEE 1547 standards series [30]. These standards have been implemented and periodically revised to deal with aspects such as voltage and power quality, grounding, islanding detection, etc [5,8,13]. The latest version, IEEE 1457-2018, was modified for inverter based microgrids. What is more, in the last 2018 release of the standard the DERs require to contain grid-supporting features such as voltage and frequency ride through capabilities and continue to function in case of parameter unbalances. This is in contrast to the original version, IEEE 1457-2003, that trip the DER offline in these situations [30–33].

In [34], the major indicators of smartness for an inverter were described. Each of these indicators contains some operational features that are described in separate sections and sub-sections in this article. Figure 1 is an illustration of the "smartness" indicators and their operational features.

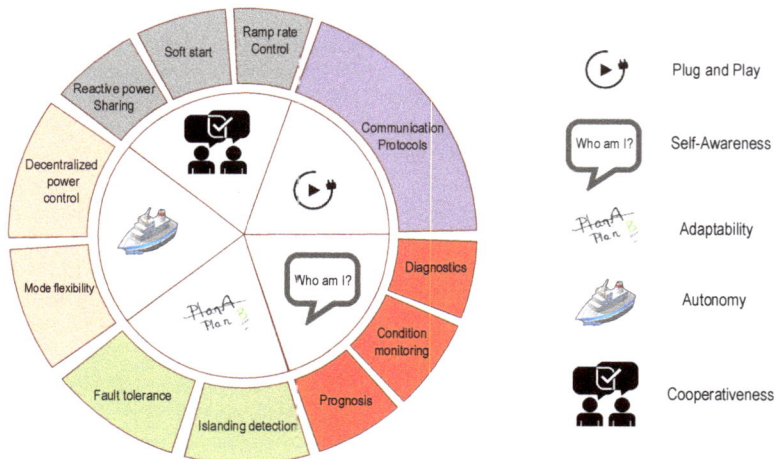

Figure 1. Indicators of smartness for smart inverters.

2.1. Plug and Play

For an inverter to be able to present this feature, it needs to be compatible with the standard communication protocols that control the microgrid. As mentioned before, the modern hierarchical control of microgrids still requires communication between different components of the grid even if they are droop based. This means that a reliable long-range communication protocol is required. In [35], the different communication technologies for smart grids has been introduced accompanied by a discussion on their advantages and disadvantages. The benefits and drawbacks of wireless connection for smart grids are presented comprehensively in [36]. Generally, the major communication requirements for a smart grid or microgrid have been presented [35,37,38] as follows:

- Data rate: This is one of the most important requirements, as not all of the communication structures are able to provide the required data rate. For example for a home or industrial area network application (HAN and IAN) the required data rate is less than 100 kbps. In contrast to a wide area application (WAN) where this figure can rise to more than 10 Mbps [39].
- Range: A microgrid consists of several DGs, energy storage systems (ESSs) and consumers. These are all interfaced to the microgrid bus by inverters that need to be able to communicate. Depending on the size of the microgrid the distance between these points can be noticeable (in the scale of tens of kilometers or more) so no all communication technologies are able to fulfill this requirement.
- Security: If the communication structure is not secure enough [40], the whole system will be vulnerable to both physical and cyber-attacks. This is a challenging aspect regarding some natural characteristics of a smart grid. For example, the smart grid is very sensitive to latency and implementing conventional security measures might introduce new and more delays [41].
- Latency: Depending on the function of the device in a smart grid, the allowed threshold for the latency is different. In the concept of inverter control signals, this allowance is less than ten milliseconds.
- Reliability and Scalability.

In order to nominate a communication protocol for smart grids and subsequently smart inverters, the aspects addressed above should be considered carefully and the differences between machine-to-machine (M2M) communication and human-to-human (H2H) communication should be understood. Both wired and wireless communication technologies can be implemented for smart inverters.

2.1.1. Wired Communication Technologies

There are some benefits to using wired communication technologies. They are more immune to interferences and their operational dependency on batteries are less than with wireless technologies. However, they present some drawbacks such as high implementation costs and signal quality degradation in power line communication (PLC) technology [35]. In fact this technology is the only wired technology with an implementation cost comparable to wireless methods, as it uses power cables for transferring data [37,42]. In addition, this technology has wide coverage, easy to implement and long range [43].

PLC technology has been divided into two categories. Narrow band (NB) PLCs, which operate in transmission frequencies bellow 500 kHz and broadband (BB) PLCs which use higher frequencies of up to 30 MHz. NB-PLC technology provides lower data rate but a much higher range. In contrast, BB-PLCs characteristics are exactly the opposite with also lower reliability. Regarding these facts, for smart inverters or smart grids the NB-PLC is a more suitable option, whereas BB-PLC can be considered for home applications [37,44,45]. In addition to PLC technologies, there are several other supports and protocols available in this category such as optical fiber based communication, Digital Subscriber Lines (DSL), coaxial cable and Ethernet [37,38]. A brief comparison of characteristics of wired communication technologies is summarized in Table 1.

Table 1. Comparison of different wired technologies for M2M communication.

Technology		Data Rate	Coverage	Disadvantages	References
PLC	NB	10–500 kbps	150 km	Difficult to achieve high bit rates. Signal attenuation. Interference from electric component connected to the line.	[37,38,46–48]
	BB	10–200 Mbps	1.5 km		
Fiber optics	PON	100 Mbps–2.5 Gbps	10–60 km	High capital costs. Difficult to upgrade.	
	AON	100 Mbps	10 km		
DSL	HDSL	2 Mbps	3.6 km	Possible data quality degradation. High operational costs.	
	ADSL	1.3–8 Mbps	5 km		
	VDSL	16–85 Mbps	1200 m		

2.1.2. Wireless Communication Technologies

In wireless technologies, one initial and obvious choice can be using the existing cellular network, as it is already available and globally dispersed. However, there are some issues regarding the different nature of M2M communication and H2H communication cellular networks are designed for the latter. Even the latest technology of cellular communication, named Long Term Evolution (LTE), is designed to cope with large data range requirements for only a small number of devices. By contrast, when investigating the characteristics of M2M communication, it can be found out that here the situation is quite different. There is a vast number of devices generating sporadic transitions of short packets, and this can overload or even shut down such a network. Furthermore, the signals and commands in M2M communication may be very sensitive to delays. Therefore, if cellular networks are to be considered as an optimal communication protocol for smart inverters they still need further improvements. As a matter of fact, the 3GPP became concerned about these issues and is currently aiming to incorporate the requirements of M2M communications in the development of the next generation of LTE or 5G [49–52].

In addition to cellular networks, there are other wireless communication technologies available. Extremely short-range technologies such as Near-field Communication (NFC) are mostly useless for inverters. Nevertheless, other short-range active radio frequency systems, such as Bluetooth or the IEEE 802.15.4 standard-based family like ZigBee and 6LoWPAN can be an alternative in some cases [49,53]. Moreover, low-power wide area networks (LPWANs) offer many of the desired features for M2M communication applications. Among the examples in this category, LoRa technology can be considered for the application to microgrids due to its low implementation and operational cost and relatively long range [49,54–56]. A comparison of these different methods is summarized in Table 2.

For M2M communication applications, the literature review highlighted ZigBee and LPWAN as the standards of choice. Comparing their specifications in Table 2, it can be observed that ZigBee provides a sufficient data rate but at the same time it has a very short range [49]. In contrast, LPWAN technologies demonstrate the opposite behavior. One of the reasons for this is the different network topologies. ZigBee systems are usually implemented following a mesh topology that provides the benefit of fault tolerance but makes the routing process more complex and not energy efficient. Furthurmore, because of the multi-hop nature of mesh network topology the actual data rate might decrease dramatically to the point that it may no longer be suitable for the application of M2M communication [49]. On the other hand, the LPWANs, or specifically LoRa and SIGFOX, networks have star topology with central nodes connecting to the internet. This topology can help with the overall simplicity of the system and less power consumption of each node. However, if a failure happens to one of the principal nodes then other nodes cannot compensate for and keep the network active [53]. Figure 2 is an illustration of the mesh (a) and star (b) topologies described before. In some countries, LPWANs technologies use cellular network bandwidths, and according to the availability and robustness of those, it can be very beneficial. Nevertheless, both short and long-range technologies have the ability to perform in unlicensed bandwidths. This is mostly an economic consideration although it may reduce the quality and security of the protocol at the same time [49,62]. Figure 3 represents a comparison of different wireless technologies based on range and data rate.

Table 2. Comparison of different wireless technologies for M2M communication.

Technology		Data Rate	Coverage	Disadvantages	References
Cellular network communication	GSM	Max 14.4 Kbps	1–10 Km	Data rates low	[35]
	GPRS	Max 170 Kbps	1–10 Km	Data rates low	
	3G	Max 2 Mbps	1–10 Km	High cost	
	WIMAX	Max 75 Mbps	Max 50 Km	Availability limited	
Short range	ZigBee	250 Kbps	Approx 50 m	Short range and low data rate	[35,49,57]
	6LoWPAN	250 Kbps	10–100 m	Short range and low data rate	[58]
	Bluetooth	1–2 Mbps	15–30 m	Short range	[49,59]
	Wi-Fi	54 Mb/s	100 m	Short range	
	UWB	110 Mb/s	10 m	Very short range	[49,57,59]
LPWAN	LoRa	0.3–37.5 Kbps	3–5 Km (Urban) 10–15 Km (Rural)	Low data range	[49,60,61]
	SIGFOX	0.1 Kbps	3–10 Km (Urban) 30–50 Km (Rural)	Low data range	
	eMTC	Less than 1 Mbps	5 km (urban) 17 km (rural)	Licensed network	

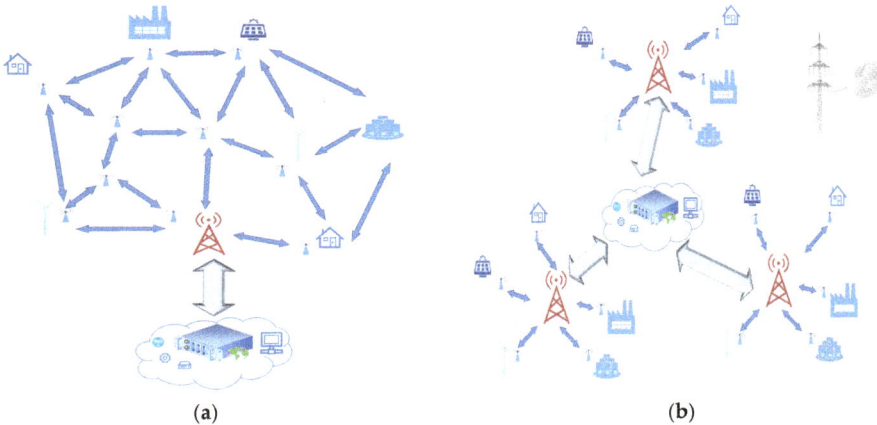

(a)   (b)

Figure 2. A schematic illustration of (**a**) the mesh topology of ZigBee and (**b**) the start topology of LoRa communication protocols.

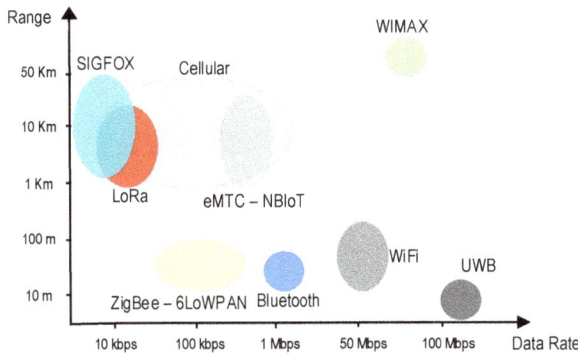

Figure 3. Comparison of different wireless technologies based on range and data rate.

2.2. Self-Awarness

In the definition of smart inverter, self-awareness is another key concept to be considered. In simple words, this means that the inverter is aware of its possible faulty parts, can determine their exact location, work out the cause of the fault, and be able to isolate them. In addition, the ability to predict possible future fault occurrences to program preventive maintenance is essential. Self-awareness can be divided into three different, yet closely related concepts [34,63,64]:

- Diagnostics so the inverter finds out the reason and origin of a fault after the occurrence.
- Condition monitoring (CM) which is a real-time evaluation of the component health status.
- Prognosis to estimate if a fault is going to happen in the future and when.

These three methods rely on fault detection procedures. For diagnostics, it is obvious that in order to work out the reason for a fault. at first, it needs to happen and be detected. For CM and prognosis, a fault or an initiation of a fault needs to be located in order to prevent the complete system to shut down or even worse. However, there are several challenges accompanied by this procedure. One problem is the effect of continuous measurements. The three concepts require measurements and it is a well-known fact that measurements cannot always be taken without interrupting the normal operation of the device. In addition, the process of deriving health indicators from physical measurements and furthermore extract prediction indicators from them can be complex [63,64].

Inverter faults are mostly triggered by thermomechanical fatigue in IGBTs [65]. They are commonly classified as, open-circuit faults, short-circuit faults and intermittent gate signal [66,67]. Detecting short-circuit faults can be easy, since in case of an occurrence, the current will rise to about four or five times the normal value. Therefore, traditional relays can be effective enough to detect and shut down the faulty component. However, the case is very different in open circuit faults, since the increase in current is not that significant, and if the sensitivity of the relays is reduced to the point they can sense these small current fluctuations, false tripping will also occur [68,69]. The intermittent behavior of RES can also cause false alarms if relays are being used for open-circuit fault detection [68]. Normally, open-circuit faults do not trigger a system shutdown but they cause the system to misbehave or work less efficient. Therefore, they can remain undetected for a long operational time resulting to other faults and further damages to the inverter [67]. As an outcome, it can be stated that open-circuit faults cannot be detected by hardware. However, there are several methods and algorithms available to effectively detect them.

In general, open circuit fault detection has been categorized into current-based methods and voltage-based methods [70]. One of the pioneering approaches in the first category is Park's vector method proposed in [71]. In this method, by applying Park vector transformation on the calculated average 3-phase current, the angle and magnitude of the currents in the complex domain are obtained, and by using pattern algorithms to analyze the trajectory of the space vector deviations, a fault can be identified. The complexity of this method is its main drawback [67,69]. In addition, this current based method and other similar ones in this category, such as two methods proposed in [72], contain another disadvantageous behavior which is a full one fundamental period delay between the occurrence and the actual detection [73]. Using voltage measurements is another criterion. Here the actual voltages at the key points of the circuit are measured and compared with their reference values derived by analytical models. This method is faster than the current based methods and the fault can be detected in one-fourth of a cycle but since voltage measurement devices are included, it adds to the complexity of the system [69,70,73].

2.3. Adaptability

It is crucial for a smart component, in this case, a smart inverter, to be able to adapt or adjust itself to the changes and occurrences of the system in which it operates. This means the ability to estimate the parameters, specifically the impedance of the grid, and self-synchronize in terms of frequency. This is adaptability and is another crucial characteristic of a smart inverter. One of the most important operational problems of microgrids is unintended islanding. This is mainly due to grid failure, and the inverters are required to be included with islanding detection algorithms in order to be able to self-adapt if required. Furthermore, being fault tolerant is another operational ability that falls in this category [34]. The importance of fault-tolerant operation becomes relevant when it is not possible to partially or fully shut down a system due to its critical application even in faulty conditions. For example, hospitals require constant and stable electrical power, and if a fault occurs the grid is required to continue the supply until the problem is solved, otherwise it can cause devastating consequences [74].

2.3.1. Fault Tolerance

As an indication of system reliability, after a fault is detected, diagnosed and isolated, the inverter and consequently the whole system are required to continue normal operation. The diagnostics and isolation procedure is crucial to prevent the problem from propagating with possibly catastrophic consequences. To be able to continue to work under faulty conditions several methods are proposed, some of them are based on the implementation of extra hardware and modification in the topology and/or the modulation procedure [74,75]. These modifications are implemented on switch level, leg level, module level or device level. At all these levels, the fault tolerance characteristic is achieved by means of redundancy. Implementation of redundant hardware at leg level is a proposed solution. In case of a switch fault, the method proposed in [76] can replace the bidirectional switch traditionally installed in a matrix converter with any of the other nine switches. The principles of the leg level criteria are simple. When the fault occurs, and it has been diagnosed the switching signals of the semiconductors of that leg are removed and transferred to the redundant leg after the bidirectional switch connecting the leg to the main circuit is triggered. The timing of these switching and replacements are also important and require precision [77,78]. In [79], the implementation of extra hardware is not included, and the methods proposed used the inherent redundancy of multi-level inverter to achieve fault tolerance characteristic, this is an example of module level. Generally, the concept of modular level fault tolerance is to bypass the faulty semiconductor switch or switches, and the resulting decreased and often unbalanced output voltage is regulated by modifying the modulation method considering a phase shift in the voltage reference to provide a balanced line to line output voltages [80,81]. The device level, which is normally implemented at industrial applications, is to use a parallel redundant inverter.

2.3.2. Islanding Detection

Unintended islanding occurrence is inevitable but must be detected as fast as possible so the distributed system can be decoupled from the grid immediately. Otherwise, it can cause harm to line workers in the main grid or deviate the voltage and frequency from the standard level. Therefore, one of the characteristics of a smart inverter is being equipped with islanding detection methods. These methods are primarily divided into remote and local detection techniques.

Remote methods are based on the detection on the consumer side and require communication between the utility and the inverter. These methods are out of the scope of this study [82–84]. Local islanding detection methods, as their name indicates, perform the detection based on local measurements. They are divided into two categories, passive and active methods. Passive methods monitor the voltage variation, frequency deviation, rate of change of power, rate of change of frequency or other trends of the system. When islanding happens these parameters will vary and if this variation exceeds a defined threshold, then islanding is detected [84]. However, if the loading of the DG does not change significantly after an islanding situation takes place, then it cannot be detected by monitoring fundamental parameters such as voltage change or frequency deviation so utilizing other parameters such as voltage unbalance or total harmonic distortion (THD) has been proposed [85]. In [86] another method which utilizes the rate of change of frequency over power has been introduced.

Active methods are based on the perturbation and later monitoring of the response of system. If unintended islanding occurs, the changes of parameters due to these perturbations will be considerable [82]. One variant of these methods is the impedance measurement method. Here, a harmonic current is injected and by monitoring the voltage response, the grid impedance can be measured. If this impedance changes for more than 0.5 Ω it can be an indication of grid failure and the DG must be decoupled. This is a fast and cheap anti-islanding method, and the problem of intentionally adding to the system disturbances can be reduced by using the harmonic signal injection at higher frequencies, 400–600 Hz [87–89].

Naturally, both passive and active methods have advantages as well as drawbacks. Passive methods are fast and do not introduce harmonics and distortion. However, they are ineffective when the load and generation are closely matched at islanded mode and generally have larger none detection zone (NDZ). By contrast, the active methods are slower, introduce harmonics to the system and, therefore, the power quality is degraded, but at the same time, they present a decent detection possibility even if there is a perfect match between load and generation in islanded mode. In other words, the active methods contain a much smaller NDZ [82–84]. Figure 4 is an illustration of islanding detecting categories with a brief summary of their advantages and disadvantages.

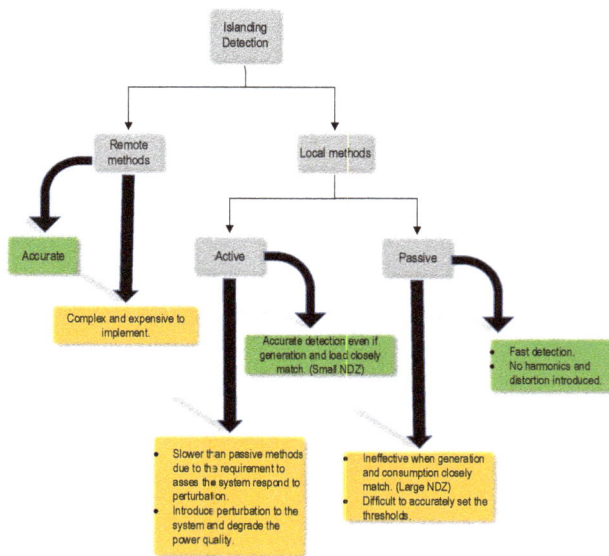

Figure 4. Islanding detection methods.

2.4. Autonomy

The different DGs and loads of a microgrid may be installed several kilometers apart, if not more. This means that the implementation of fast communication links to control every aspect of their operation is neither feasible nor cost-effective. On the other hand, the requirement of communication in microgrids cannot be completely omitted, but the inverters need to present some autonomous characteristics to be able to function properly. In this section, the most important autonomous features for smart inverters has been addressed. Control of the power flow using local measurements based on droop methods is explained briefly, as it is the foundation of the decentralized control idea. In addition, the mode flexibility feature is described. The importance of this feature emerges from the fact that it is necessary for a smart inverter to be able to autonomously switch operating modes due to different microgrid configurations.

2.4.1. Decentralized Power Control

An inverter, as an interface for a DG, is required to be equipped with control schemes to control its active and reactive power flow autonomously. One method is using droop control schemes, which can control active and reactive power sharing effectively for parallel inverters only by using local measurements. These methods have been conventionally used for large power systems based on synchronous generators and gradually became a well-established method for microgrids. Droop control methods and more comprehensive proposed structures based on these methods, such as hierarchical control, have been extensively reviewed and presented in the literature [8,9,12–14,16,17, 19,28,29,90–99]. By implementing droop control accurate power sharing between parallel connected inverters can be achieved with regard to their power ratings. In an AC microgrid P-ω and Q-V droops are being used for this purpose. The active and reactive output power can be calculated by local voltage-current measurements from low-pass filters. Then by using the droop Equations (1) and (2) and using the no-load values for ω_0 and V_0, grid values for V and ω can be worked out:

$$\omega = \omega_0 - mP, \tag{1}$$

$$V = V_0 - nQ, \tag{2}$$

It shall be noted that in Equations (1) and (2) P and Q are the active and reactive delivered powers of each inverter, so the total power consumed by the loads is $P = P_1 + P_2 + \cdots$ and the same for reactive power. m and n are droop coefficients, and in simple droop, they are calculated in a way that they take the power rating of each component into effect:

$$m_1 S_1 = m_2 S_2 = \cdots, \tag{3}$$

$$n_1 S_1 = n_2 S_2 = \cdots, \tag{4}$$

In Equations (3) and (4) S values are the apparent power ratings of each DGs. By solving Equations (1) and (2) the reference voltage can be extracted that will be used by the inverters inner control loops to regulate its power flow. In other words, the droop control provides references for the inner control loops of the inverter. In more complex control schemes, which consist of more than one layer of control, the secondary control compensates for the frequency and voltage deviation resulted from the droop control. This means that it transmits voltage and frequency references for the primary level (droop). The droop coefficients are also influenced by energy management consideration of higher levels of control, *tertiary* level of hierarchical control [13]. Figure 5 illustrates the droop control concept for active power including the secondary control effect.

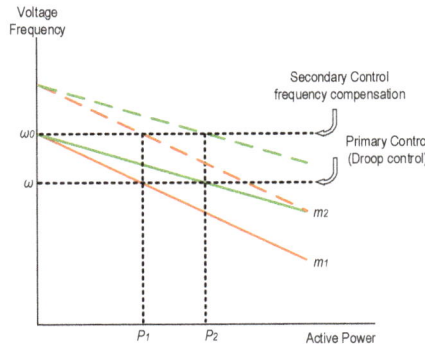

Figure 5. Illustration of active power sharing between two parallel inverters using conventional droop method. The secondary response of the hierarchical control is also included.

2.4.2. Mode Flexibility

Another autonomous feature for an inverter is its ability to change its operational mode without external intervention in a smooth and seamless way. Depending on the control structure, an inverter can operate as grid forming, grid feeding and grid supporting [10]. In grid-connected mode, since the parameters are set by the grid the inverters perform as either grid feeding or grid supporting. By contrast, in islanded mode, there must be at least one grid forming inverter setting the voltage and frequency references of the grid. Figure 6 is a representation of these different configurations. As it can be observed in Figure 6a, the grid forming inverter functions as an ideal voltage source with fixed voltage and frequency. The additional loop (in green) is a descriptive illustration of voltage base grid supporting inverter or VSI. The control structure of these inverters is designed to work out the active and reactive power flow of each inverter and by using different control schemes provide voltage and frequency references. Figure 6b presents the same concept for grid feeding inverters and CSIs, which inject fixed or variable active and reactive power according to their inner loops and MPPT algorithms if they are of renewable energy nature [100].

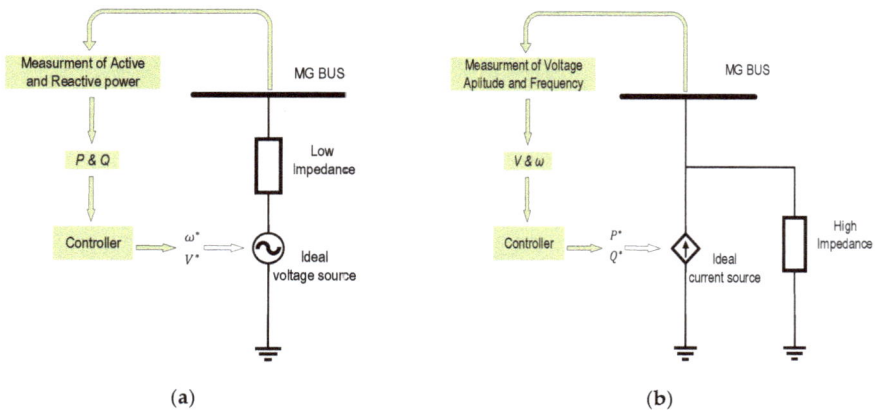

Figure 6. Representation of different modes of inverters in microgrids. (**a**), grid forming/grid supporting VSI, (**b**), grid feeding/grid supporting CSI.

The concern here is the transition state when inverters roles are switched between these different modes. This switching is required due to the transition of the state of the microgrid from grid connected to islanded or the other way around. One side effect of this phenomenon is frequency unbalances and voltage and current disturbances so the inverters are required to be equipped with suitable control schemes the be able to compensate for these transient effects and provide a smooth transient state [94,101–104].

In [103] a method has been proposed consisting of a phase-locked-loop (PLL) in order to retrieve the inverters output voltage angle and use it as a reference for droop control. In addition, a virtual inductor is used to limit the current inrush. Therefore, a smooth transition mode can be achieved. There are other approaches using a similar idea, based on PLL controllers. For instance, in [105] a method with the capability of ensuring a smooth transition in booth direction, grid connected to islanded and islanded to grid connected, has been proposed. Another method gathers voltage samples from grid and point of common coupling (PCC) at islanded mode, the values can be fed to a synchronization controller and after that been introduced to the droop control scheme, as reference values, which is regulating voltage and frequency at islanded mode. In this manner, a seamless mode transition is achieved after synchronization [106].

2.5. Cooperativeness

An inverter, as a part of a grid, is required to function according to the functionality of other components of the grid. This is cooperativeness. One of the major aspects in this area is power sharing, which has been addressed before by introducing hierarchical droop based controls. However, by implementing the conventional droop, although active power sharing can be sufficiently achieved by using local measurements, reactive power sharing is not that accurate and also the problem of harmonics current sharing is not addressed. Furthermore, there are other aspects such as ramp rate control for renewable energy sources and soft start capability, which lie in this category.

2.5.1. Reactive Power and Harmonic Current Sharing

The problem of reactive power sharing inaccuracy in conventional droop methods is derived from the fact that it depends on the impedance of DG feeders and out-put side impedance of L-C-L filters. This contrasts with the active power sharing control, which is always accurate. In order to overcome this problem, several techniques have been proposed. The first group of methods are based on introducing a virtual impedance loop in order to compensate for the inverter output and line impedance unbalances and therefore improving the reactive power sharing. Several approaches based on this idea can be found in the literature, and their sole drawback is due to the possible large value of the virtual impedance. This may degrade the voltage quality particularly in weak islanded microgrids [19,107–109]. The concept of virtual impedance loop can be described by the following equations:

$$\vartheta_0 = \vartheta_0^* - i_0 Z_0(s), \tag{5}$$

where ϑ_0 is the voltage reference provided for inner control loops, ϑ_0^* is the voltage reference calculated by the droop control loop and $Z_0(s)$ is the virtual impedance transfer function. Same as the m and n droop coefficient calculations in Equations (3) and (4), Z_0 values for different units are calculated according to their power ratings:

$$Z_{01} S_1 = Z_{02} S_2 = \cdots, \tag{6}$$

Considering Equations (5) and (6) it can be understood that the virtual impedance implementation method generates a new voltage reference for inner control loops of the inverter by modifying the initial references set by the droop algorithms. Many approaches are conducted to empower the virtual impedance shaping technique. By using communication, the feeder impedance can be actively compensated, and this improves the reactive power sharing quality [8,13]. Modifying the droop method is another method to improve reactive current sharing Over the years, many improved techniques have been proposed and implemented [110,111]. In [110], an enhanced droop method has been proposed. In this method, the reactive power sharing is improved by the sharing error reduction operation. However, this procedure will cause an output voltage reduction which is regulated separately. Signal injection techniques can be mentioned as an example of other methods proposed for this purpose [112].

The sharing of harmonic currents generated because of the presence of non-linear loads, such as rectifiers, have not been addressed by conventional droop methods [107,113–115]. This presence of harmonics can affect the voltage and current wave-forms, cause overheating problems and add to losses, and lead to stability problems, therefore needed to be shared accurately [116,117]. In [108] a modification in the virtual impedance loop was proposed. This method provides the resistive output impedance for higher order current harmonics by subtracting the voltages corresponding to current harmonics from the reference voltage:

$$\vartheta_0 = \vartheta_0^* - i_{01}.sL_D - \sum_{\substack{h=3 \\ odd}}^{11} i_{oh} R_h, \tag{7}$$

where i_{01} is the output current, L_D is the virtual impedance and R_h is the resistive coefficient corresponding to each of the harmonic terms, i_{oh} By using Equation (7), a new voltage reference is generated for the inner loops, covering effects of both linear and non-linear loads.

2.5.2. Soft Start

When a DG is initiating its energy generation, its interface inverter will try to start injecting power into the microgrid. If this procedure occurs without proper initiations, it will cause transient disturbances in grid parameters which are obviously not favorable. The ability of soft start for inverters can be achieved by modifying the virtual impedance loop. The soft starter controls the virtual impedance and is programmed to assign it with a higher value at the beginning of the transient state and then gradually reduce it to its nominal value. This is a modification for the virtual impedance loop mentioned before as part of the power sharing control (specifically to improve the reactive power and harmonic current sharing). Therefore as the control structure is already equipped with a virtual impedance loop it is just required to include the soft starter [94,99,118,119]. The implementation of the soft starter is done by the following Equation (8):

$$Z_D(t) = Z_f - \left(Z_f - Z_i\right)e^{\frac{-t}{T}},\tag{8}$$

where Z_f and Z_i are the initial and final values of the virtual impedance and T is the time interval of the soft start procedure. By using this method, the system can start with a large impedance and gradually decrease it to its designed value for normal operation [13,120,121].

2.5.3. Ramp Rate Control

DGs in microgrids are mostly of a renewable energy nature and it is a well-known fact that these type of energies, specifically wind and solar, suffer fluctuations in their power generation due to climate-caused effects. These power ramp rates are required to be controlled in a defined boundary in order to reduce the risk of possible damage occurrences when for example the clouds suddenly clear out over a large solar power station. The concept of ramp rate control is simple and can be represented by a simple equation for a solar panel [122,123] as in Equation (9):

$$R(t) = \frac{dP_{PV}}{dt} = \frac{\Delta P}{\Delta T},\tag{9}$$

where $R(t)$ is the instant ramp rate, and P_{PV} is the instant power generated by the panel. When the ramp rate exceeds a defined value, the output power requires to be regulated. Incorporating energy storage units such as batteries, flywheels and supercapacitors or reliable energy generation units such as fuel cells are the common methods to compensate for these fluctuations and smoothen the generation [124–127]. Using Equation (10), the energy storage capacity required for ramp rate control according to the power rating/perimeter scale of a solar field and the maximum acceptable ramp rate value can be worked out [128,129]:

$$C_{ESS} = \frac{1.8\,P^*}{3600}\left[\frac{90}{2\,R_{max}} - \tau\right],\tag{10}$$

$$\tau = a\cdot l + b,\tag{11}$$

where C_{ESS} is the capacity of the energy storage unit, R_{max} is the maximum ramp rate and P^* is the power of the solar field. Equation (11) represents the time constant τ, which is an empirical correlation with the shortest measured length of the solar field perimeter, l. ($a = 0.042$ s/m and $b = -0.5$ s). In [124] implementation of electric double layer capacitor (EDLC) for smoothening the generation of PV panels alongside with the size consideration of the capacitor and the utilization of a fuel cell to compensate for slow generation fluctuations have been discussed. Although most of the ramp rate control methods are based on using energy storage devices, there are other possibilities. For example, In [122], by modifying the MPPT algorithms, which renewable energy systems functions are based on, the ramp rate can be controlled by deviating from the set point of the MPPT when required, to balance the output power.

3. Discussion

In order to define a device as smart, it has to represent several qualities and characteristics mostly depending on its application and implementation. In this study, these characteristics have been described for inverters, which are used as interfaces for DGs in microgrids:

- Plug-and-play capability has been the first concept discussed. Regardless of how much effort has been expended to omit the communication requirement, in order to achieve a stable and functional microgrid, especially in the context of accurate active, reactive and harmonic current sharing, this requirement still exists. In contrast to other smartness indicators that evaluate the smartness based on how less it relies on communication for normal operation, plug-and-play is mainly related to communication compatibility. Different relevant communication technologies have been introduced and investigated in detail with tabular information. The fact that the DGs can be spread over a large area, maybe several kilometers apart, will emphasize the infeasibility of using conventional communication protocols mostly according to economic considerations. One practical and attractive solution can be the use of the Internet. However, in accordance with M2M communication and the Internet of things concept, it is not practical to connect every node of a system directly to the web. A long range, low implementation and operating cost, protocol that is reliable enough at the same time, such as LoRa or ZigBee, can be used to communicate each of the inverters to internet gateways. It is understood, further research and thorough experimental work are required to end up with a conclusion and chose one communication protocol over another.

- Self-awareness is another smartness indicator. In this section, the concept of fault diagnosis, condition monitoring and prognosis have been explained. In addition, fault detection methods have been briefly reviewed since all three-health indicators of inverters require fault detection at some point in their control structure. According to various methods proposed for open-circuit fault detection, it can be observed that, no matter how complicated the detection algorithm could be, there are still based on local measurements with the aim of avoiding the further requirement of communication. Short-circuit fault detection can be effectively conducted by use of hardware and local current measurements.

- Another studied characteristic of smart inverters has been their adaptability. This is related to the change of parameters of the grid, the loads and the working mode of the inverter itself. Here smartness is the ability to sense and identify some fluctuations in the parameters readings that can be an indication for faults. Then, effectively self-adjust to be able to continue to function regardless of the mode change or the fault as it is not always possible to shut down the system immediately even if a serious problem has been detected. This is called fault tolerance, and the different methods have been introduced and compared. Some of them require extra, redundant, hardware and others provide redundancy by manipulating the modulation process. Unintended islanding for microgrids due to grid failure is another issue, and if the system cannot self-adapt itself with this phenomenon, the results can be catastrophic. Methods for islanding detection are different, and they have been briefly introduced and compared. It is understood, same as the fault detection methods, in islanding detection methods the aim of reducing communication requirements and self-adapting in the procedure is a major consideration nowadays.

- The other important characteristic is autonomy, which intends to reduce the requirements for communications among inverters installed far away from each other even more. In microgrids, depending on if it is islanded or grid-connected, the inverters are controlled as VSCs or CSCs connected in parallel. One measure of autonomy is the ability to control the active and reactive power sharing among them with minimum required communication. The droop control methods empower this characteristic by controlling each DGs power flow according to its rated values, by mostly using local measurements to minimizing communication requirements. Another aspect that has been considered here is when the working mode of the microgrid changes from islanded to grid-connected or vice versa. Naturally, in that case, the role of most inverters is required to be changed as well, since in islanded mode the inverters are mostly controlled as grid forming, in contrast to a grid-connected mode when they transform to grid feeding or grid supporting modes. These transformations are required to be done automatically, and each of the inverters should be equipped with suitable control structures to make these mode switching accompanied by unavoidable transient responses seamless and smooth.

- The last concept covered has been cooperativeness. By definition, this means the smartness that an inverter requires to be able to function in accordance and alongside with other inverters in a grid. All the inverters are required to take some responsibility to regulate and compensate for unbalances and disturbances present in the system. In addition, their operation and behavior should be in alignment with other neighboring components. Otherwise, further disturbances will be introduced to the system. Active, reactive and harmonics current sharing is the most important characteristic fitting in this definition. In contrast to some of the characteristics discussed before the requirement of communication can be sensed for cooperativeness.

As a brief summary of the concepts addressed above it can be stated that smartness of an inverter in a microgrid, mostly refers to minimizing communication requirements for its normal operation. However, according to the current literature, this requirement is still present and cannot be omitted altogether. Currently, different communication protocols are used for this application, as well as other M2M communication applications, but each of them have their own drawbacks and are far from perfect. Therefore, there is a broad area of research and development available to fill this gap. One possible idea is the use of the Internet infrastructure which sounds perfect according to its worldwide availability and its great storage and computational potentials based on the cloud concept. Nevertheless, here, the most important consideration is not the internet platform itself. On the contrary, the main aspect is to propose an architecture to connect the components to the cloud. Such an architecture is required to present these qualities:

- Reliable; no data can be lost or misinterpreted.
- Fast, thresholds for delays are very limited.
- Secure, immune from unregistered intervention.
- Globally available and compatible.
- Low energy consumption.
- Robustness.
- Bidirectional.
- Economically feasible, in terms of implementation, operational costs and maintenance.

Containing these characteristics, a structure based on a wireless communication protocol can be proposed for inverters communication to improve microgrids performance and control. In spite of that, it should always be remembered that every communication structure, no matter how carefully engineered and well established it is, still contain deficiencies such as delays, by nature. Therefore, the research and development of control structures centered on the goal of reducing the communication requirement should continue further more.

Author Contributions: B.A.-Z. carried out the literature review and wrote the paper; E.J.P.-G. supervised and provided critical reviews. J.C.V. and J.M.G. coordinated and reviewed the final work.

Funding: This research was funded by the Aalborg University Talent Programme 2016 with the research project: The Energy Internet–Integrating Internet of Things into the Smart Grid.

Acknowledgments: This work was funded by the AAU Talent Project: The Energy Internet–Integrating Internet of Things into the Smart Grid (http://www.iot-energy.et.aau.dk).

Conflicts of Interest: The authors declare no conflict of interest.

References

1. Colson, C.M.; Nehrir, M.H. A review of challenges to real-time power management of microgrids. In Proceedings of the 2009 IEEE Power & Energy Society General Meeting, Calgary, AB, Canada, 26–30 July 2009; pp. 1–8.
2. Pepermans, G.; Driesen, J.; Haeseldonckx, D.; Belmans, R.; D'haeseleer, W. Distributed generation: Definition, benefits and issues. *Energy Policy* **2005**, *33*, 787–798. [CrossRef]
3. Lasseter, R.; Piagi, P. Microgrid: A Conceptual Solution. In Proceedings of the 2004 IEEE 35th Annual Power Electronics Specialists Conference, Aachen, Germany, 20–25 June 2004.
4. Marnay, C.; Chatzivasileiadis, S.; Abbey, C.; Iravani, R.; Joos, G.; Lombardi, P.; Mancarella, P.; Von Appen, J. Microgrid evolution roadmap. In Proceedings of the 2015 International Symposium on Smart Electric Distribution Systems and Technologies, Vienna, Austria, 8–11 September 2015; pp. 139–144.
5. Carrasco, J.M.; Garcia Franquelo, L.; Bialasiewicz, J.T.; Galván, E.; Portillo Guisado, R.C.; Martín Prats, M.D.L.Á.; León, J.I.; Moreno-Alfonso, N. Power-Electronic Systems for the Grid Integration of Renewable Energy Sources: A Survey. *IEEE Trans. Ind. Electron.* **2004**, *53*, 1002–1016. [CrossRef]
6. Prodanović, M.; Green, T.C. Control and filter design of three-phase inverters for high power quality grid connection. *IEEE Trans. Power Electron.* **2003**, *18*, 373–380. [CrossRef]
7. Zhong, Q.C.; Weiss, G. Synchronverters: Inverters that mimic synchronous generators. *IEEE Trans. Ind. Electron.* **2011**, *58*, 1259–1267. [CrossRef]
8. Guerrero, J.M.; Chandorkar, M.; Lee, T.L.; Loh, P.C. Advanced control architectures for intelligent microgridspart i: Decentralized and hierarchical control. *IEEE Trans. Ind. Electron.* **2013**, *60*, 1254–1262. [CrossRef]
9. Guerrero, J.M.; Loh, P.C.; Lee, T.L.; Chandorkar, M. Advanced control architectures for intelligent microgridsPart II: Power quality, energy storage, and AC/DC microgrids. *IEEE Trans. Ind. Electron.* **2013**, *60*, 1263–1270. [CrossRef]
10. Rocabert, J.; Luna, A.; Blaabjerg, F.; Rodríguez, P. Control of power converters in AC microgrids. *IEEE Trans. Power Electron.* **2012**, *27*, 4734–4749. [CrossRef]
11. Kim, J.; Guerrero, J.M.; Rodriguez, P.; Teodorescu, R.; Nam, K. Mode adaptive droop control with virtual output impedances for an inverter-based flexible AC microgrid. *IEEE Trans. Power Electron.* **2011**, *26*, 689–701. [CrossRef]
12. Lu, X.; Guerrero, J.M.; Sun, K.; Vasquez, J.C. An improved droop control method for dc microgrids based on low bandwidth communication with dc bus voltage restoration and enhanced current sharing accuracy. *IEEE Trans. Power Electron.* **2014**, *29*, 1800–1812. [CrossRef]
13. Guerrero, J.M.; Vasquez, J.C.; Matas, J.; De Vicuña, L.G.; Castilla, M. Hierarchical control of droop-controlled AC and DC microgrids—A general approach toward standardization. *IEEE Trans. Ind. Electron.* **2011**, *58*, 158–172. [CrossRef]
14. Yang, N.; Paire, D.; Gao, F.; Miraoui, A.; Liu, W. Compensation of droop control using common load condition in DC microgrids to improve voltage regulation and load sharing. *Int. J. Electr. Power Energy Syst.* **2015**, *64*, 752–760. [CrossRef]
15. Alsafran, A. Literature Review of Power Sharing Control Strategies in Islanded AC Microgrids with Nonlinear Loads. In Proceedings of the 2018 IEEE PES Innovative Smart Grid Technologies Conference Europe (ISGT-Europe), Sarajevo, Bosnia-Herzegovina, 21–25 October 2018; pp. 1–6.
16. Guerrero, J.M.; de Vicuna, L.G.; Matas, J.; Castilla, M.; Miret, J. A wireless controller to enhance dynamic performance of parallel inverters in distributed generation systems. *IEEE Trans. Power Electron.* **2004**, *19*, 1205–1213. [CrossRef]
17. Kim, J.W.; Choi, H.S.; Cho, B.H. A novel droop method for converter parallel operation. *IEEE Trans. Power Electron.* **2002**, *17*, 25–32.
18. Mihalache, L. Paralleling control technique with no intercommunication signals for resonant controller-based inverters. In Proceedings of the 38th IAS Annual Meeting on Conference Record of the Industry Applications Conference, Salt Lake City, UT, USA, 12–16 October 2003; Volume 3, pp. 1882–1889.

19. Guerrero, J.M.; de Vicuna, L.; Matas, J.; Castilla, M.; Miret, J. Output Impedance Design of Parallel-Connected {UPS} Inverters With Wireless Load-Sharing Control. *IEEE Trans. Ind. Electron.* **2005**, *52*, 1126–1135. [CrossRef]
20. Converters, T.; Borup, U.; Blaabjerg, F.; Member, S.; Enjeti, P.N. Sharing of Nonlinear Load in Parallel-Connected. *IEEE Trans. Ind. Appl.* **2001**, *37*, 1817–1823.
21. Chang, J.M. Parallel operation of series-connected PWM voltage regulators without control interconnection. *Proc. IEEE-Electr. Power Appl.* **2001**, *148*, 141–147. [CrossRef]
22. Chiang, S.J.; Yen, C.Y.; Chang, K.T. A Multimodule Parallelable Series-Connected. *IEEE Trans. Ind. Electron.* **2001**, *48*, 506–516. [CrossRef]
23. Song, S.H.; Kang, S.I.; Hahm, N.K. Implementation and control of grid connected AC-DC-AC power converter for variable speed wind energy conversion system. In Proceedings of the Eighteenth Annual IEEE Applied Power Electronics Conference and Exposition, Miami Beach, FL, USA, 9–13 February 2003; pp. 154–158.
24. Kim, J.S.; Sul, S.K. New control scheme for AC-DC-AC converter without DC link electrolytic capacitor. In Proceedings of the IEEE Power Electronics Specialist Conference, Seattle, WA, USA, 20–24 June 1993; pp. 300–306.
25. Kakigano, H.; Miura, Y.; Uchida, R.; Engineering, I. DC Micro-grid for Super High Quality Distribution. *IEEE Power Electron. Spec. Conf.* **2010**, *25*, 3066–3075. [CrossRef]
26. Salomonsson, D.; Soder, L.; Sannino, A. An Adaptive Control System for a DC Microgrid for Data Centers. *IEEE Trans. Ind. Appl.* **2008**, *44*, 1910–1917. [CrossRef]
27. Che, L.; Shahidehpour, M. DC microgrids: Economic operation and enhancement of resilience by hierarchical control. *IEEE Trans. Smart Grid* **2014**, *5*, 2517–2526.
28. Bidram, A.; Davoudi, A. Hierarchical structure of microgrids control system. *IEEE Trans. Smart Grid* **2012**, *3*, 1963–1976. [CrossRef]
29. Nutkani, I.U.; Loh, P.C.; Wang, P.; Blaabjerg, F. Autonomous droop scheme with reduced generation cost. *IEEE Trans. Ind. Electron.* **2014**, *61*, 6803–6811. [CrossRef]
30. Photovoltaics, D.G.; Storage, E. IEEE Standard for Interconnection and Interoperability of Distributed Energy Resources with Associated Electric Power Systems Interfaces. In *IEEE Std 1547-2018 (Revision IEEE Std 1547-2003)*; IEEE: Piscataway, NJ, USA, 2018; pp. 1–138.
31. Hoke, A.; Giraldez, J.; Palmintier, B.; Ifuku, E.; Asano, M.; Ueda, R.; Symko-Davies, M. Setting the Smart Solar Standard: Collaborations Between Hawaiian Electric and the National Renewable Energy Laboratory. *IEEE Power Energy Mag.* **2018**, *16*, 18–29. [CrossRef]
32. Mahmud, R.; Hoke, A.; Narang, D. Validating the test procedures described in UL 1741 SA and IEEE. In Proceedings of the 2018 IEEE 7th World Conference on Photovoltaic Energy Conversion (WCPEC), Waikoloa Village, HI, USA, 10–15 June 2018; pp. 1445–1450.
33. Behravesh, V.; Keypour, R.; Foroud, A.A. Stochastic analysis of solar and wind hybrid rooftop generation systems and their impact on voltage behavior in low voltage distribution systems. *Sol. Energy* **2018**, *166*, 317–333. [CrossRef]
34. Guerrero, J.M.; Xue, Y. Smart Inverters for Utility and Industry Applications. In Proceedings of the PCIM Europe 2015 International Exhibition and Conference for Power Electronics, Intelligent Motion, Renewable Energy and Energy Management, Nuremberg, Germany, 19–20 May 2015; pp. 277–284.
35. Supriya, S.; Magheshwari, M.; Sree Udhyalakshmi, S.; Subhashini, R. Musthafa Smart grid technologies: Communication technologies and standards. *Int. J. Appl. Eng. Res.* **2015**, *10*, 16932–16941.
36. Gungor, V.C.; Lu, B.; Hancke, G.P. Opportunities and challenges of wireless sensor networks in smart grid. *IEEE Trans. Ind. Electron.* **2010**, *57*, 3557–3564. [CrossRef]
37. Ancillotti, E.; Bruno, R.; Conti, M. The role of communication systems in smart grids: Architectures, technical solutions and research challenges. *Comput. Commun.* **2013**, *36*, 1665–1697. [CrossRef]
38. Kuzlu, M.; Pipattanasomporn, M.; Rahman, S. Communication network requirements for major smart grid applications in HAN, NAN and WAN. *Comput. Netw.* **2014**, *67*, 74–88. [CrossRef]
39. Kuzlu, M.; Pipattanasomporn, M.; Tech, V. Assessment of Communication Technologies and Network Requirements for Different Smart Grid Applications. In Proceedings of the IEEE PES Innovative Smart Grid Technologies Conference (ISGT), Washington, DC, USA, 24–27 February 2013.
40. Tankard, C. The security issues of the Internet of Things. *Comput. Fraud Secur.* **2015**, *2015*, 11–14. [CrossRef]

41. Bekara, C. Security issues and challenges for the IoT-based smart grid. *Procedia Comput. Sci.* **2014**, *34*, 532–537. [CrossRef]

42. Galli, S.; Scaglione, A.; Wang, Z. For the grid and through the grid: The role of power line communications in the smart grid. *Proc. IEEE* **2011**, *99*, 998–1027. [CrossRef]

43. Yigit, M.; Gungor, V.C.; Tuna, G.; Rangoussi, M.; Fadel, E. Power line communication technologies for smart grid applications: A review of advances and challenges. *Comput. Netw.* **2014**, *70*, 366–383. [CrossRef]

44. Usman, A.; Shami, S.H. Evolution of communication technologies for smart grid applications. *Renew. Sustain. Energy Rev.* **2013**, *19*, 191–199. [CrossRef]

45. Galli, S.; Scaglione, A.; Wang, Z. Power Line Communications and the Smart Grid. In Proceedings of the 2010 First IEEE International Conference on Smart Grid Communications, Gaithersburg, MD, USA, 4–6 October 2010; pp. 303–308.

46. Kabalci, Y. A survey on smart metering and smart grid communication. *Renew. Sustain. Energy Rev.* **2016**, *57*, 302–318. [CrossRef]

47. Faheem, M.; Shah, S.B.H.; Butt, R.A.; Raza, B.; Anwar, M.; Ashraf, M.W.; Ngadi, M.A.; Gungor, V.C. Smart grid communication and information technologies in the perspective of Industry 4.0: Opportunities and challenges. *Comput. Sci. Rev.* **2018**, *30*, 1–30. [CrossRef]

48. Shaukat, N.; Ali, S.M.; Mehmood, C.A.; Khan, B.; Jawad, M.; Farid, U.; Ullah, Z.; Anwar, S.M.; Majid, M. A survey on consumers empowerment, communication technologies, and renewable generation penetration within Smart Grid. *Renew. Sustain. Energy Rev.* **2018**, *81*, 1453–1475. [CrossRef]

49. Centenaro, M.; Vangelista, L.; Zanella, A.; Zorzi, M. Long-range communications in unlicensed bands: The rising stars in the IoT and smart city scenarios. *IEEE Wirel. Commun.* **2016**, *23*, 60–67. [CrossRef]

50. Laya, A.; Alonso, L.; Alonso-Zarate, J. Is the random access channel of LTE and LTE-A suitable for M2M communications? A survey of alternatives. *IEEE Commun. Surv. Tutor.* **2014**, *16*, 4–16. [CrossRef]

51. Biral, A.; Centenaro, M.; Zanella, A.; Vangelista, L.; Zorzi, M. The challenges of M2M massive access in wireless cellular networks. *Digit. Commun. Netw.* **2015**, *1*, 1–19. [CrossRef]

52. Lien, S.Y.; Chen, K.C.; Lin, Y. Toward ubiquitous massive accesses in 3GPP machine-to-machine communications. *IEEE Commun. Mag.* **2011**, *49*, 66–74. [CrossRef]

53. Townsend, C.; Arms, S. Wireless sensor networks: Principles and aplications. In *Sensor Technology Handbook*; Wilson, J.S., Ed.; Elsevier: Amsterdam, The Netherlands, 2005; pp. 575–589.

54. Terashmila, L.K.A.; Iqbal, T.; Mann, G. A comparison of low cost wireless communication methods for remote control of grid-tied converters. In Proceedings of the Canadian Conference on Electrical and Computer Engineering, Windsor, ON, Canada, 30 April–3 May 2017; pp. 1–4.

55. Thielemans, S.; Bezunartea, M.; Steenhaut, K. Establishing transparent IPv6 communication on LoRa based low power wide area networks (LPWANS). In Proceedings of the 2017 Wireless Telecommunications Symposium (WTS), Chicago, IL, USA, 26–28 April 2017; pp. 1–6.

56. Angrisani, L.; Arpaia, P.; Bonavolonta, F.; Conti, M.; Liccardo, A. LoRa protocol performance assessment in critical noise conditions. In Proceedings of the 2017 IEEE 3rd International Forum on Research and Technologies for Society and Industry (RTSI), Modena, Italy, 11–13 September 2017.

57. Lee, J.; Su, Y.; Shen, C. A Comparative Study of Wireless Protocols: Bluetooth, UWB, ZigBee, and Wi-Fi. In Proceedings of the IECON 2007—33rd Annual Conference of the IEEE Industrial Electronics Society, Taipei, Taiwan, 5–8 November 2007; pp. 46–51.

58. Chen, Y.; Hou, K.M.; Zhou, H.; Shi, H.L.; Liu, X.; Diao, X.; Ding, H.; Li, J.J.; De Vaulx, C. 6LoWPAN stacks: A survey. In Proceedings of the 2011 7th International Conference on Wireless Communications, Networking and Mobile Computing, Wuhan, China, 23–25 September 2011; pp. 1–4.

59. Ferro, E.; Potortì, F. Bluetooth and Wi-Fi wireless protocols: A survey and a comparison. *IEEE Wirel. Commun.* **2005**, *12*, 12–26. [CrossRef]

60. Chen, J.; Hu, K.; Wang, Q.; Sun, Y.; Shi, Z.; He, S. Narrowband Internet of Things: Implementations and Applications. *IEEE Internet Things J.* **2017**, *4*, 2309–2314. [CrossRef]

61. Petäjäjärvi, J.; Mikhaylov, K.; Roivainen, A.; Hänninen, T.; Pettissalo, M. On the coverage of LPWANs: Range evaluation and channel attenuation model for LoRa technology. In Proceedings of the 2015 14th International Conference on ITS Telecommunications (ITST), Copenhagen, Denmark, 2–4 December 2016; pp. 55–59.

62. Bardyn, J.P.; Melly, T.; Seller, O.; Sornin, N. IoT: The era of LPWAN is starting now. In Proceedings of the ESSCIRC Conference 2016: 42nd European Solid-State Circuits Conference, Lausanne, Switzerland, 12–15 September 2016; pp. 25–30.

63. Shaoyong, Y.; Dawei, X.; Angus, B.; Philip, M.; Li, R.; Peter, T. Condition Monitoring for Device Reliability in Power Electronic Converters: A Review. *IEEE Trans. Power Electron.* **2010**, *25*, 2734–2752.

64. Oh, H.; Han, B.; McCluskey, P.; Han, C.; Youn, B.D. Physics-of-failure, condition monitoring, and prognostics of insulated gate bipolar transistor modules: A review. *IEEE Trans. Power Electron.* **2015**, *30*, 2413–2426. [CrossRef]

65. Lamb, J.; Mirafzal, B. Open-Circuit IGBT Fault Detection and Location Isolation for Cascaded Multilevel Converters. *IEEE Trans. Ind. Electron.* **2017**, *64*, 4846–4856. [CrossRef]

66. Lu, B.; Sharma, S.K. A literature review of IGBT fault diagnostic and protection methods for power inverters. *IEEE Trans. Ind. Appl.* **2009**, *45*, 1770–1777.

67. Estima, J.O.; Cardoso, A.J.M. A new approach for real-time multiple open-circuit fault diagnosis in voltage-source inverters. *IEEE Trans. Ind. Appl.* **2011**, *47*, 2487–2494. [CrossRef]

68. Zeineldin, H.H.; Kan'an, N.H.; Casagrande, E.; Woon, W.L. Data mining approach to fault detection for isolated inverter-based microgrids. *IET Gener. Transm. Distrib.* **2013**, *7*, 745–754.

69. Moosavi, S.S.; Kazemi, A.; Akbari, H. A comparison of various open-circuit fault detection methods in the IGBT-based DC/AC inverter used in electric vehicle. *Eng. Fail. Anal.* **2019**, *96*, 223–235. [CrossRef]

70. An, Q.T.; Sun, L.Z.; Sun, L.; Jahns, T.M. Low-cost diagnostic method for open-switch faults in inverters. *Electron. Lett.* **2010**, *46*, 1021. [CrossRef]

71. Mendes, A.M.S.; Cardoso, A.J.M. Voltage Source Inverter Fault in Variable Speed Ac Current Approach. In Proceedings of the IEEE International Electric Machines and Drives Conference, Seattle, WA, USA, 9–12 May 1999; pp. 704–706.

72. Peuget, R.; Courtine, S.; Rognon, J.P. Fault detection and isolation on a pwm inverter. *IEEE Trans. Ind. Appl.* **1997**, *34*, 1471–1478.

73. De Araujo Ribeiro, R.L.; Jacobina, C.B.; Da Silva, E.R.C.; Lima, A.M.N. Fault detection of open-switch damage in voltage-fed PWM motor drive systems. *IEEE Trans. Power Electron.* **2003**, *18*, 587–593. [CrossRef]

74. Zhang, W.; Xu, D.; Enjeti, P.N.; Li, H.; Hawke, J.T.; Krishnamoorthy, H.S. Survey on fault-tolerant techniques for power electronic converters. *IEEE Trans. Power Electron.* **2014**, *29*, 6319–6331. [CrossRef]

75. Lezana, P.; Pou, J.; Meynard, T.A.; Rodriguez, J.; Ceballos, S.; Richardeau, F. Survey on fault operation on multilevel inverters. *IEEE Trans. Ind. Electron.* **2010**, *57*, 2207–2218. [CrossRef]

76. Andreu, J.; Kortabarria, I.; Ibarra, E.; Martínez De Alegría, I.; Robles, E. A new hardware solution for a fault tolerant matrix converter. In Proceedings of the 2009 35th Annual Conference of IEEE Industrial Electronics, Porto, Portugal, 3–5 November 2009; pp. 4469–4474.

77. Weber, P.; Poure, P.; Theilliol, D.; Saadate, S. Design of hardware fault tolerant control architecture for Wind Energy Conversion System with DFIG based on reliability analysis. In Proceedings of the 2008 IEEE International Symposium on Industrial Electronics, Cambridge, UK, 30 June–2 July 2008; pp. 2323–2328.

78. Rodríguez, M.A.; Claudio, A.; Theilliol, D.; Vela, L.G.; Hernández, L. Strategy to replace the damaged power device for fault-tolerant induction motor drive. In Proceedings of the 2009 Twenty-Fourth Annual IEEE Applied Power Electronics Conference and Exposition, Washington, DC, USA, 15–19 February 2009; Volume 52, pp. 343–346.

79. Li, S.; Xu, L. Strategies of fault tolerant operation for three-level PWM inverters. *IEEE Trans. Power Electron.* **2006**, *21*, 933–940. [CrossRef]

80. Yi, Z.; Hongge, S.; Bin, X. Optimization of neutral shift in cell-fault treatment for cascaded H-bridge. In Proceedings of the 2008 International Conference on Electrical Machines and Systems, Wuhan, China, 17–20 October 2008; pp. 1683–1685.

81. Correa, P.; Pacas, M.; Rodríguez, J. Modulation strategies for fault-tolerant operation of H-bridge multilevel inverters. *IEEE Int. Symp. Ind. Electron.* **2006**, *2*, 1589–1594.

82. Mahat, P.; Chen, Z.; Bak-Jensen, B. Review of Islanding Detection Methods for Distributed Generation. In Proceedings of the 2008 Third International Conference on Electric Utility Deregulation and Restructuring and Power Technologies, Nanjing, China, 6–9 April 2008; Volume 88, pp. 2743–2748.

83. De Mango, F.; Liserre, M.; Dell'Aquila, A.; Pigazo, A. Overview of anti-islanding algorithms for PV systems. Part I: Passive methods. In Proceedings of the 2006 12th International Power Electronics and Motion Control Conference, Portoroz, Slovenia, 30 August–1 September 2007; pp. 1–6.

84. Li, C.; Cao, C.; Cao, Y.; Kuang, Y.; Zeng, L.; Fang, B. A review of islanding detection methods for microgrid. *Renew. Sustain. Energy Rev.* **2014**, *35*, 211–220. [CrossRef]

85. Kim, K.-H.; Jang, S.-I. An Islanding Detection Method for Distributed Generations Using Voltage Unbalance and Total Harmonic Distortion of Current. *IEEE Trans. Power Deliv.* **2004**, *19*, 745–752.

86. Pai, F.-S.; Huang, S.-J. A detection algorithm for islanding-prevention of dispersed consumer-owned storage and generating units. *IEEE Trans. Energy Convers.* **2001**, *16*, 346–351. [CrossRef]

87. Ciobotaru, M.; Teodorescu, R.; Blaabjerg, F. On-line grid impedance estimation based on harmonic injection for grid-connected PV inverter. In Proceedings of the 2007 IEEE International Symposium on Industrial Electronics, Vigo, Spain, 4–7 June 2007; pp. 2437–2442.

88. Asiminoaei, L.; Teodorescu, R.; Blaabjerg, F.; Borup, U. A digital controlled PV-inverter with grid impedance estimation for ENS detection. *IEEE Trans. Power Electron.* **2005**, *20*, 1480–1490. [CrossRef]

89. Asiminoaei, L.; Teodorescu, R.; Blaabjerg, F.; Borup, U. A new method of on-line grid impedance estimation for PV inverter. In Proceedings of the Nineteenth Annual IEEE Applied Power Electronics Conference and Exposition, Anaheim, CA, USA, 22–26 February 2004; pp. 1527–1533.

90. Hua, C.; Liao, K.; Lin, J.; City, T.; Country, Y. Parallel Operation of Inverters for Distributed Photovoltaic Power Supply System. In Proceedings of the 2002 IEEE 33rd Annual IEEE Power Electronics Specialists Conference, Cairns, Australia, 23–27 June 2002; pp. 1979–1983.

91. Simpson-Porco, J.W.; Dörfler, F.; Bullo, F. Synchronization and power sharing for droop-controlled inverters in islanded microgrids. *Automatica* **2013**, *49*, 2603–2611. [CrossRef]

92. Tuladha, A.; Jin, H.; Unger, T.; Mauch, K. Control of parrallel invertes in distributed AC power systems with considerartion of the line impedance effect. *IEEE Trans. Ind. Appl.* **1998**, *36*, 321–328.

93. Vasquez, J.C.; Guerrero, J.M.; Savaghebi, M.; Eloy-Garcia, J.; Teodorescu, R. Modeling, analysis, and design of stationary-reference-frame droop-controlled parallel three-phase voltage source inverters. *IEEE Trans. Ind. Electron.* **2013**, *60*, 1271–1280. [CrossRef]

94. Guerrero, J.M.; Vásquez, J.C.; Matas, J.; Castilla, M.; García de Vicuna, L. Control strategy for flexible microgrid based on parallel line-interactive UPS systems. *IEEE Trans. Ind. Electron.* **2009**, *56*, 726–736. [CrossRef]

95. Yu, X.; Khambadkone, A.M.; Wang, H.; Terence, S.T.S. Control of parallel-connected power converters for low-voltage microgrid—Part I: A hybrid control architecture. *IEEE Trans. Power Electron.* **2010**, *25*, 2962–2970. [CrossRef]

96. Yao, W.; Chen, M.; Matas, J.; Guerrero, J.M.; Qian, Z.M. Design and analysis of the droop control method for parallel inverters considering the impact of the complex impedance on the power sharing. *IEEE Trans. Ind. Electron.* **2011**, *58*, 576–588. [CrossRef]

97. Mohamed, Y.A.R.I.; El-Saadany, E.F. Adaptive decentralized droop controller to preserve power sharing stability of paralleled inverters in distributed generation microgrids. *IEEE Trans. Power Electron.* **2008**, *23*, 2806–2816. [CrossRef]

98. Zhong, Q.C. Robust droop controller for accurate proportional load sharing among inverters operated in parallel. *IEEE Trans. Ind. Electron.* **2013**, *60*, 1281–1290. [CrossRef]

99. Guerrero, J.M.; Matas, J.; De Vicuña, L.G.; Castilla, M.; Miret, J. Wireless-control strategy for parallel operation of distributed-generation inverters. *IEEE Trans. Ind. Electron.* **2006**, *53*, 1461–1470. [CrossRef]

100. Monshizadeh, P.; De Persis, C.; Monshizadeh, N.; Van Der Schaft, A. A communication-free master-slave microgrid with power sharing. In Proceedings of the 2016 American Control Conference (ACC), Boston, MA, USA, 6–8 July 2016; pp. 3564–3569.

101. Wang, J.; Chang, N.C.P.; Feng, X.; Monti, A. Design of a Generalized Control Algorithm for Parallel Inverters for Smooth Microgrid Transition Operation. *IEEE Trans. Ind. Electron.* **2015**, *62*, 4900–4914. [CrossRef]

102. Katiraei, F.; Iravani, M.R.; Lehn, P.W. Micro-grid autonomous operation during and subsequent to islanding process. *IEEE Trans. Power Deliv.* **2005**, *20*, 248–257. [CrossRef]

103. Hu, S.H.; Kuo, C.Y.; Lee, T.L.; Guerrero, J.M. Droop-controlled inverters with seamless transition between islanding and grid-connected operations. In Proceedings of the 2011 IEEE Energy Conversion Congress and Exposition, Phoenix, AZ, USA, 17–22 September 2011; pp. 2196–2201.

104. Vandoorn, T.L.; Meersman, B.; De Kooning, J.D.M.; Vandevelde, L. Transition from islanded to grid-connected mode of microgrids with voltage-based droop control. *IEEE Trans. Power Syst.* **2013**, *28*, 2545–2553. [CrossRef]

105. Tran, T.V.; Chun, T.W.; Lee, H.H.; Kim, H.G.; Nho, E.C. PLL-based seamless transfer control between grid-connected and islanding modes in grid-connected inverters. *IEEE Trans. Power Electron.* **2014**, *29*, 5218–5228. [CrossRef]

106. Jin, C.; Gao, M.; Lv, X.; Chen, M. A seamless transfer strategy of islanded and grid-connected mode switching for microgrid based on droop control. In Proceedings of the 2012 IEEE Energy Conversion Congress and Exposition (ECCE), Raleigh, NC, USA, 15–20 September 2012; pp. 969–973.

107. He, J.; Li, Y.W.; Blaabjerg, F. An enhanced islanding microgrid reactive power, imbalance power, and harmonic power sharing scheme. *IEEE Trans. Power Electron.* **2015**, *30*, 3389–3401. [CrossRef]

108. Guerrero, J.M.M.; Matas, J.; Garcia de Vicuna, L.; Berbel, N.; Sosa, J. Wireless-Control Strategy for Parallel Operation of Distributed Generation Inverters. In Proceedings of the IEEE International Symposium on Industrial Electronics, Dubrovnik, Croatia, 20–23 June 2005; pp. 845–850.

109. Axelrod, B.; Berkovich, Y.; Ioinovici, A. Virtual Impedance Loop for Droop-Controlled Single-Phase Parallel Inverters Using a Second-Order. In Proceedings of the 2003 International Symposium on Circuits and Systems, Bangkok, Thailand, 25–28 May 2003; pp. 2993–3002.

110. Han, H.; Liu, Y.; Sun, Y.; Su, M.; Guerrero, J.M. An improved droop control strategy for reactive power sharing in islanded microgrid. *IEEE Trans. Power Electron.* **2015**, *30*, 3133–3141. [CrossRef]

111. Yaoqin, J.; Dingkun, L.; Shengkui, P. Improved droop control of parallel inverter system in standalone microgrid. In Proceedings of the 8th International Conference on Power Electronics—ECCE Asia, Jeju, Korea, 30 May–3 June 2011; pp. 1506–1513.

112. He, J.; Li, Y.W. An enhanced microgrid load demand sharing strategy. *IEEE Trans. Power Electron.* **2012**, *27*, 3984–3995. [CrossRef]

113. Vandoorn, T.; Meersman, B.; De Kooning, J.; Vandevelde, L. Controllable harmonic current sharing in islanded microgrids: DG units with programmable resistive behavior toward harmonics. *IEEE Trans. Power Deliv.* **2012**, *27*, 831–841. [CrossRef]

114. Sreekumar, P.; Khadkikar, V. Nonlinear load sharing in low voltage microgrid using negative virtual harmonic impedance. In Proceedings of the IECON 2015—41st Annual Conference of the IEEE Industrial Electronics Society, Yokohama, Japan, 9–12 November 2015; pp. 3353–3358.

115. Lorzadeh, I.; Abyaneh, H.A.; Savaghebi, M.; Guerrero, J.M. A hierarchical control scheme for reactive power and harmonic current sharing in islanded microgrids. In Proceedings of the 2015 17th European Conference on Power Electronics and Applications (EPE'15 ECCE-Europe), Geneva, Switzerland, 8–10 September 2015; pp. 1–10.

116. Moussa, H.; Shahin, A.; Martin, J.P.; Nahid-Mobarakeh, B.; Pierfederici, S.; Moubayed, N. Harmonic Power Sharing with Voltage Distortion Compensation of Droop Controlled Islanded Microgrids. *IEEE Trans. Smart Grid* **2018**, *9*, 5335–5347. [CrossRef]

117. Micallef, A.; Apap, M.; Spiteri-Staines, C.; Guerrero, J.M. Mitigation of Harmonics in Grid-Connected and Islanded Microgrids Via Virtual Admittances and Impedances. *IEEE Trans. Smart Grid* **2017**, *8*, 651–661. [CrossRef]

118. Guerrero, J.M.; Berbel, N.; Matas, J.; Sosa, J.L.; De Vicuña, L.G. Control of line-interactive UPS connected in parallel forming a microgrid. In Proceedings of the IEEE International Symposium on Industrial Electronics, Vigo, Spain, 4–7 June 2007; pp. 2667–2672.

119. Chen, Y.; Wang, Z.; Zhou, X.; Zhou, L.; Chen, Z.; Luo, A.; Wang, M. Seamless transfer control strategy for three-phase inverter in microgrid. In Proceedings of the 2016 IEEE 8th International Power Electronics and Motion Control Conference (IPEMC-ECCE Asia), Hefei, China, 22–26 May 2016; pp. 1759–1763.

120. Guerrerol, J.M.; Berbel, N.; Matas, J.; Sosa, J.L.; De Vicuña, L.G. Droop Control Method with Virtual Output Impedance for Parallel Operation of Uninterruptible Power Supply Systems in a Microgrid. In Proceedings of the APEC 07—Twenty-Second Annual IEEE Applied Power Electronics Conference and Exposition, Anaheim, CA, USA, 25 February–1 March 2007.

121. Guerrero, J.M.; Hang, L.; Uceda, J. Control of Distributed Uninterruptible Power Supply Systems. *IEEE Trans. Ind. Electron.* **2008**, *55*, 2845–2859. [CrossRef]

122. Sangwongwanich, A.; Yang, Y.; Blaabjerg, F. A cost-effective power ramp-rate control strategy for single-phase two-stage grid-connected photovoltaic systems. In Proceedings of the 2016 IEEE Energy Conversion Congress and Exposition (ECCE), Milwaukee, WI, USA, 18–22 September 2016; pp. 1–7.

123. Salehi, V.; Radibratovic, B. Ramp rate control of photovoltaic power plant output using energy storage devices. In Proceedings of the 2014 IEEE PES General Meeting | Conference & Exposition, National Harbor, MD, USA, 27–31 July 2014; pp. 1–5.

124. Kakimoto, N.; Satoh, H.; Takayama, S.; Nakamura, K. Ramp-rate control of photovoltaic generator with electric double-layer capacitor. *IEEE Trans. Energy Convers.* **2009**, *24*, 465–473. [CrossRef]

125. Van Haaren, R.; Morjaria, M.; Fthenakis, V. An energy storage algorithm for ramp rate control of utility scale PV (photovoltaics) plants. *Energy* **2015**, *91*, 894–902. [CrossRef]

126. Kasem, A.H.; El-Saadany, E.F.; El-Tamaly, H.H.; Wahab, M.A.A. Power ramp rate control and flicker mitigation for directly grid connected wind turbines. *IET Renew. Power Gener.* **2010**, *4*, 261. [CrossRef]

127. Cormode, D.; Cronin, A.D.; Richardson, W.; Lorenzo, A.T.; Brooks, A.E.; Dellagiustina, D.N. Comparing ramp rates from large and small PV systems, and selection of batteries for ramp rate control. In Proceedings of the 2013 IEEE 39th Photovoltaic Specialists Conference (PVSC), Tampa, FL, USA, 16–21 June 2013; pp. 1805–1810.

128. De la Parra, I.; Marcos, J.; García, M.; Marroyo, L. Control strategies to use the minimum energy storage requirement for PV power ramp-rate control. *Sol. Energy* **2015**, *111*, 332–343. [CrossRef]

129. Marcos, J.; Storkël, O.; Marroyo, L.; Garcia, M.; Lorenzo, E. Storage requirements for PV power ramp-rate control. *Sol. Energy* **2014**, *99*, 28–35. [CrossRef]

energies

MDPI

Article

An Analysis of Frequency Variations and its Implications on Connected Equipment for a Nanogrid during Islanded Operation

Jakob Nömm *, Sarah K. Rönnberg * and Math H. J. Bollen *

Electric Power Engineering, Luleå University of Technology, 931 87 Skellefteå, Sweden
* Correspondence: jakob.nomm@ltu.se (J.N.); sarah.ronnberg@ltu.se (S.K.R.); math.bollen@ltu.se (M.H.J.B.)

Received: 29 July 2018; Accepted: 12 September 2018; Published: 16 September 2018

Abstract: Frequency, voltage and reliability data have been collected in a nanogrid for 48 weeks during islanded operation. Frequency values from the 48 week measurements were analyzed and compared to relevant limits. During 19.5% of the 48 weeks, the nanogrid had curtailed the production due to insufficient consumption in islanded operation. The curtailment of production was also the main cause of the frequency variations above the limits. When the microgrid operated on stored battery power, the frequency variations were less than in the Swedish national grid. 39.4% of all the interruptions that occurred in the nanogrid are also indirectly caused by the curtailment of solar production. Possible solutions for mitigating the frequency variations and lowering the number of interruptions are also discussed.

Keywords: frequency variations; islanded operation; nanogrids; power quality; power system reliability

1. Introduction

Microgrids and nanogrids are potential solutions for providing better electrical service for areas that are insufficiently served by the traditional electricity grid. The same microgrids and nanogrids could also provide economic and environmental benefits in remote areas [1]. The term nanogrid has been suggested for defining a small microgrid [2], for instance a single house. Nanogrids can operate in either grid connected mode or in islanded mode.

Long term measurements of power quality indices for a nanogrid during islanded operation are needed in order to evaluate the performance and long term effects on the connected equipment in the nanogrid. This paper presents 48 weeks of frequency, voltage and reliability data for a nanogrid during islanded operation, including transitions between grid operation and islanded operation. The presented data is the main contribution of the paper.

The frequency measurements are compared to European electrical standards EN 50160 [3] and EN 50160/A1 [4] in order to establish if the frequency variations in the nanogrid during islanded operation surpasses the range set for systems without synchronous connection to an interconnected system. The frequency variations are also compared to International Electrotechnical Commission (IEC) Standard 60034-1 [5], computer power supply ATX12V design specifications [6], Intel power supply design specifications [7] and IEC Standard 60076-1 [8] to predict possible effects on connected equipment. The frequency variations are also correlated to the number of interruptions and the total downtime that occurred during the measured 48 weeks.

1.1. The Nanogrid

The nanogrid where the measurements have been collected is located in the southern part of Sweden. It has a 20 kWp photovoltaic installation on the roof and a 2.6 kWp photovoltaic installation on the facade. The nanogrid has a 144 kWh lead acid battery and 1100 kWh hydrogen storage.

The operating topology is a 3-phase 50 Hz system with 230 V RMS phase-to-neutral voltage. The solar-battery-hydrogen system is intended to be the primary energy system where a backup 15 kVA diesel generator is intended to be the secondary energy system that starts if the primary energy system fails. If the failure is prolonged or if the diesel generator does not start, the nanogrid will then connect to the low voltage utility grid. One example of when the nanogrid connects to the utility grid is if the available energy in the primary and secondary energy system is not sufficient to supply the consumption. The basic energy system overview for the nanogrid can be seen in Figure 1.

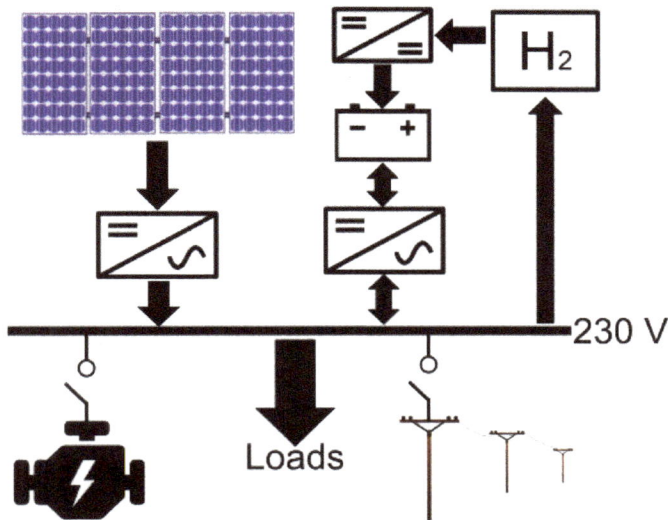

Figure 1. Basic overview of the nanogrid energy system.

During island operation, the nanogrid is operated by nine different inverters, two SMA Solar Technology Sunny Tripower inverters are used for two separate 10 kWp photovoltaic installations on the roof, one SMA Sunny Boy inverter is used for a 2.6 kWp photovoltaic installations on the facade and six SMA Sunny Island inverters are used for control of the battery charging and discharging. One of the SMA Sunny Island inverters also controls the electrolysis of water for the production of hydrogen and the hydrogen fuel cell to convert hydrogen to electricity. The consumption in the house consists of normal household appliances, a 3-phase heat pump, two electric cars and the electrolyzer to produce hydrogen.

By using the battery and hydrogen storage, a certain amount of the produced solar power is lost due to the conversion losses. However, some of the conversion losses are used to heat the house during the winter. The diesel generator operated for 43 h in the measured 48 weeks of islanded operation and delivered 473 kWh to the loads in the nanogrid. The yearly electricity consumption for the house is around 17,000 kWh and 4000 kWh for the two electric cars. For more information regarding the nanogrid see [9].

1.2. Frequency Control in the Nanogrid

The frequency in the nanogrid is controlled by the SMA Sunny Island inverters which uses frequency-shift power control (FSPC) [10] and SMA Automatic Frequency Adjustment (AFA) [11]. The FSPC is used to keep the balance between load and generation. During sunny days with not enough consumption, the FSPC increases the frequency to above 51 Hz to signal the SMA Tripower solar inverters that production should be curtailed. The amount of curtailment increases linearly between 51 and 52 Hz from 0 to 100%. The FSPC uses the battery voltage to determine the appropriate frequency in the islanded nanogrid depending on the amount of load that is present.

Another feature of the FSPC is the shutdown of the solar inverters by increasing the frequency towards 55 Hz. This is done in order for the Sunny Island inverters to synchronize to an external source, which for this nanogrid is the utility grid. The AFA compensates for the over frequency by temporarily shifting the frequency to 49 Hz to enable clocks to run at the correct time. This correction occurs on a 12 h basis [11].

2. Results

2.1. Frequency Variations during Island Operation

The 10 s average values of the frequency for the 48 weeks when the nanogrid operated in islanded operation was used to create an empirical cumulative distribution function. The results are shown in the top part of Figure 2. For 19.5% of the 48 week period, the nanogrid is not utilizing the entire solar power production and the FSPC decreases power output. In the figure, this is when the frequency exceeds 51 Hz. The 49 Hz frequency value that stands for about 30.7% of the 48 week period is caused by the AFA compensating for the over frequency. The Cumulative Distribution Function (CDF) for 54 weeks of 10 s average frequency measurements when the nanogrid was in grid-connected operation mode is presented in the lower part of Figure 2 for a comparison with the nanogrid islanded operation.

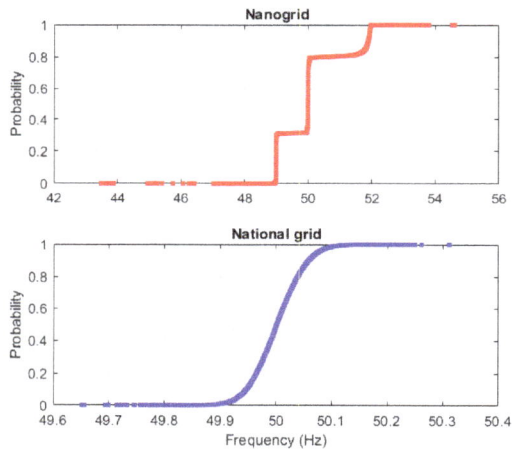

Figure 2. CDF for the 48 weeks of 10 s average frequency values in the nanogrid during islanded operation (**top**) and a corresponding CDF (**bottom**) for the 54 weeks of 10 s average frequency values in the nanogrid during grid operation. Note the difference in horizontal scale.

One typical frequency regulation scenario is when there is not enough consumption (including battery charging and electrolysis of water) during the day when the solar PV installation is producing power. When this happens, the FSPC increases the frequency above 51 Hz to curtail the production. During the night the AFA shifts the frequency to a lower value than 50 Hz in order to compensate the time increase for clocks. One example of this scenario can be observed in Figure 3 where the 10 s average frequency is plotted for a 34.5 h period. The plot starts at 01:00 the 15th of April 2017 and ends the 16th of April 2017 at 11:30. At the 15th between 01:00 and about 06:03 at sunrise, there is insufficient solar production and the load is drawing power from the battery storage and therefore the load is matched to the source giving a frequency value near 50 Hz. Between 06:03 and 11:13 the solar production together with the battery storage is matched to the load which gives a frequency value near 50 Hz, but with some variations that can be observed more clearly in the top part of Figure 4. From

11:13 to 18:03 the FSPC curtails the solar production in order to match to the load. The regulation starts at 11:13 at about 60% curtailment and increases to around 90% at 14:27. Between 18:03 the 15th April and 06:48 the 16th April the load is served mainly by the battery but since there has been a substantial amount of over frequency during the day, the AFA compensates for the over frequency by operating the nanogrid at 49 Hz. Between 06:48 and 11:30 the 16th the load is served by the solar generation and battery storage, since the generation is matched by the load the frequency value is 50 Hz.

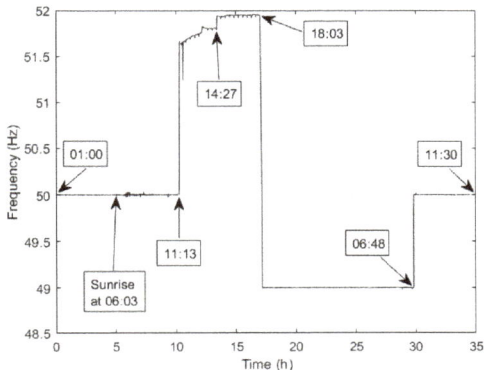

Figure 3. Frequency variations in the nanogrid from April 15th 2017 01:00 to 11:30 the 16th of April 2017.

When the source is matched to the load, the frequency is much closer to 50 Hz than in the Swedish national grid. This can be observed in Figure 4 where the top part of the plot shows a zoomed in view of Figure 3. It can be seen that the frequency starts to vary more just after sunrise when the solar production starts to increase. The total load in the nanogrid is varying between 0.7 to 5.4 kW in the duration shown in Figure 4.

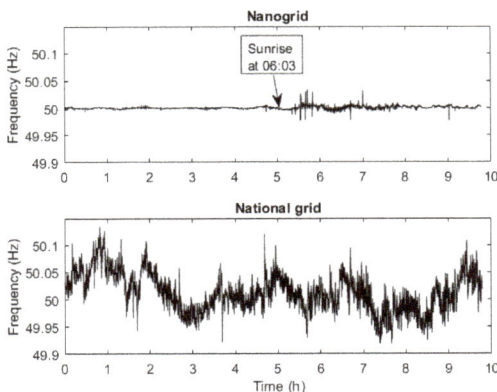

Figure 4. Enlarged view of the frequency variations in the nanogrid from Figure 3 (**top**) and frequency in the Swedish national grid (**bottom**) during the same period of time.

2.2. Minimum and Maximum Values Observed

The highest frequency values of 55 Hz occur when there is not enough loads to consume the entire production from the PV installation. This happens when the batteries are fully charged, the hydrogen tank is full and the consumption in the house is low. An example of this can be observed in the top part of Figure 5. The plot starts at 08:00 and ends at 19:00 the 25th September 2016. For this occasion, the majority of the 55 Hz values occur approximately every 23 min.

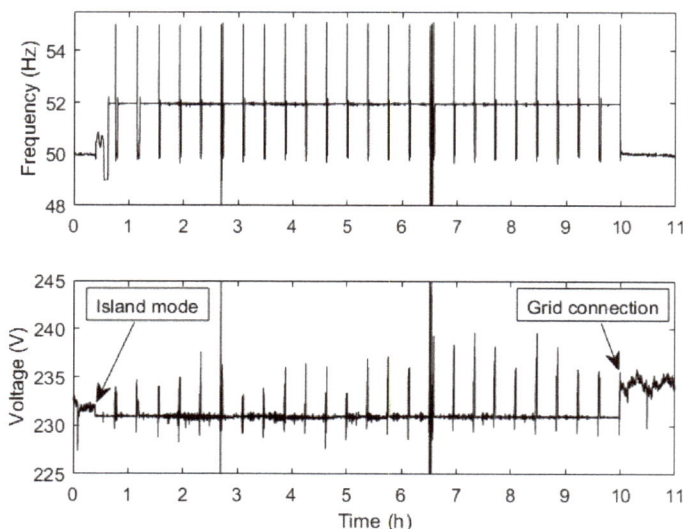

Figure 5. The one cycle average frequency between 08:00 and 19:00 the 25th September 2016 when there is not enough loads to consume the entire production from the PV installation (**top**) with the corresponding one cycle average phase-to-neutral voltage for one phase (**bottom**). The frequency scale is truncated at 48 Hz since there are two interruptions in the plot that make the frequency drop to zero. The voltage scale is truncated at 245 V and 225 V to give a better representation of the voltage variations.

When the frequency reaches about 55 Hz the nanogrid connects to the utility grid for about 10 to 50 s after which the Sunny Island inverters switch back to island operation. Occasionally, the transition causes a short interruption, in this case at h 2:40 (10:40 real time) and 6:30 (14:30 real time) in Figure 5. For the most part, these transitions only cause a rise in the phase-to-neutral voltage of a few volts for the duration of the grid connection which can be seen in the lower part of Figure 5.

Examples of the lowest frequency values that occurred can be seen in the top part of Figure 6. The lowest frequency values occur just after an interruption with duration of 0.9 s when the frequency reached 55 Hz about 10 to 50 s earlier. At this occasion, the intended grid connection failed and the nanogrid experienced an interruption.

After the short interruption, the Sunny Island inverters power up again in islanded operation with a frequency that starts at a value of about 44 Hz that shortly decreases towards 41 to 42 Hz which could then drop below 40 Hz for a one or two cycles. The frequency then gradually increases towards 49 Hz and then increase towards 52 Hz.

One of the phase-to-neutral voltages before and after the two interruptions can be seen in the lower part of Figure 6 where the left interruption is seen more clearly in Figure 7. During 4 s after the voltage recovers from the interruption, the RMS voltage fluctuates with a peak to peak magnitude of 15.2 to 35.5 V RMS at a frequency of about 12.5 to 16.7 Hz. This frequency range is in the 3 to 33 Hz span in which the eye is most sensitive to flicker [12]. This voltage fluctuation range and frequency will cause flickering of incandescent lamps. However, in this nanogrid only LED lamps are used which could be more or less sensitive to the voltage variation magnitude and frequency in terms of flicker output [13]. The largest and lowest 1 s average frequency value in the 48 week measurements when the nanogrid operated in islanded operation were 55.2 Hz and 41.3 Hz, respectively.

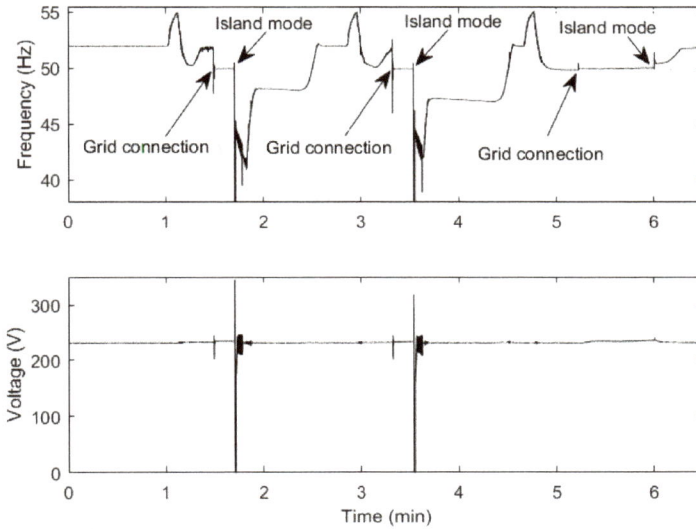

Figure 6. Enlarged view of the largest frequency variations from Figure 5 (**top**) and its corresponding voltage (**bottom**) for one phase. The scale is truncated at 38 Hz since the frequency drops towards 0 during the interruptions.

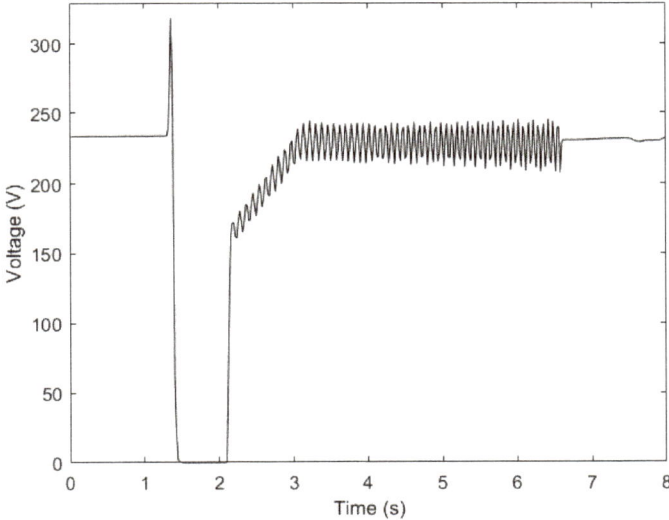

Figure 7. Enlarged view of the one cycle RMS voltage for the left interruption in Figure 6.

2.3. Comparison to Standards

According to European Standard EN 50160, the 10 s average frequency should remain between 49 and 51 Hz for 95% of one week and should remain between 42.5 and 57.5 Hz for 100% of the time for systems without synchronous connection to an interconnected system. The 95% confidence interval for the islanded operation CDF in Figure 2 spans 48.99 to 51.95 Hz. The lowest 10 s average value was 43.48 Hz and the highest was 54.61 Hz.

The grid connected frequency values in Figure 2 always remain within the specified frequency range for interconnected systems, according to EN 50160. The islanded operation data divided into weekly sections is shown in Figure 8 where the vertical axis is the amount of 10 s average values outside of the range 49 to 51 Hz every week.

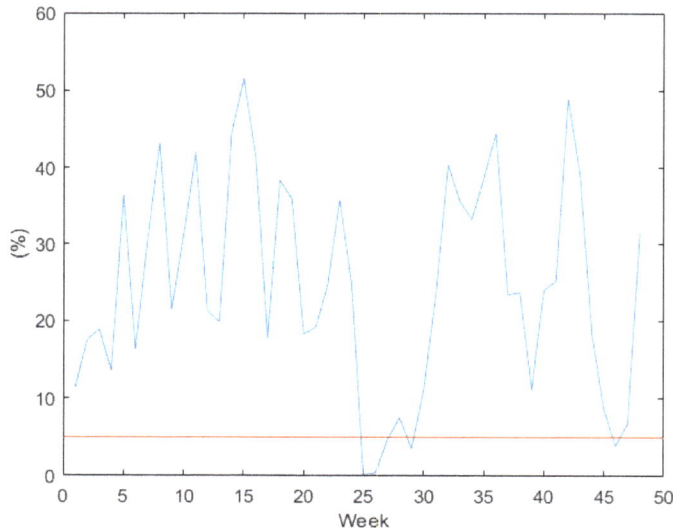

Figure 8. The amount of 10 s average values of the frequency per week outside the range 49 to 51 Hz. The straight line represents the 5% limit in EN 50160.

The islanded operation data is an assembly of long and short periods of time where the nanogrid is operating in islanded operation. This means that not all of the measurements are continuous and therefore Figure 8 does not have a definitive correlation with the seasons of the year. The straight line in Figure 8 is the 5% weekly limit of allowed values that can exceed the range 49 to 51 Hz.

In total, 89.6% of the 48 week measurements do not fall in the range that EN 50160 has set for systems without synchronous connection to an interconnected system. However, all the 10 s average frequency values remain within the maximum allowed variation set by EN 50160 for systems without synchronous connection to an interconnected system which is 42.5 to 57.5 Hz. The 95% confidence interval for the grid measurements in Figure 2 spans 49.92 to 50.09 Hz. The nanogrid is within this range for 48.3% of the 48 week measurements. This means that one could expect almost half of the time to have the same frequency quality in the nanogrid as what is normally seen in the Swedish national grid.

If the nanogrid were to be located in the neighboring country Norway, the requirement in Annex EN 50160/A1 would apply. This document states that for systems without synchronous connection to an interconnected system, the frequency shall remain within 49 to 51 Hz for 100% of the time. With this requirement, the total probability of being outside the frequency range set by EN 50160/A1 is 25% of the 48 week period. The reason to why the EN 50160/A1 Standard has a larger acceptance number is due to the fact that the entire 48 week period is considered and not individual weeks. However, the maximum allowed variations from the rated frequency is surpassed in the EN 50160/A1 Standard, but not for the EN 50160 Standard. This is since the EN 50160 Standard allows a larger frequency span of 42.5 to 57.5 Hz while the EN 50160/A1 Standard allows a frequency of 49 to 51 Hz. However, any load connected would still see the same frequency variation, regardless of which standard would apply. A summary of Section 2.3 can be seen in Table 1.

Table 1. Summary of Section 2.3 for the nanogrid during islanded operation.

Variable	EN 50160	EN 50160/A1
10 s average frequency limit (100% of the time)	42.5 to 57.5 Hz	49 to 51 Hz
10 s average frequency limit (95% of the time per week)	49 to 51 Hz	No such limit
Time outside limit (% of total measured time)	89.6	25

3. Frequency Variations and their Effects on Equipment

In general, the frequency variations in a large interconnected grid are small, so the impact on different types of equipment is almost negligible [14]. However, as seen in the previous sections, the frequency variations within an island-operated nanogrid are larger than in a large interconnected grid. The nanogrid in this case is a residential house and not an industrial facility that could need a precise frequency for the correct operation of the facility. The question arises of whether the frequency variations between about 41.3 and 55.2 Hz will have a large negative impact on household appliances. Universal motors that are used in for instance portable tools can be run on any input frequency and will therefore not be affected by the frequency variations [15]. Induction motors that drive household equipment like refrigerators and heat pumps will run at different speeds depending on the frequency. A large increase in V/f ratio will cause saturation of the induction motor and therefore the induction motor could get overheated due to higher currents being drawn. IEC Standard 60034-1 [5] defines two zones of operation for electrical AC motors. The first zone is Zone A in which the motor operation should not be affected by the variations in voltage and frequency except from a slight increase in operating temperature. The second zone is Zone B where operation should be avoided in occurrence, time and magnitude.

If operation in Zone B takes place often or continuously the motor should be de-rated to fit those operating conditions. The 10 min average values for the phase-to-phase voltage with the corresponding frequency for the 48 week island operated measurements are shown in Figure 9. The 10 min average 54 week grid connected measurements are also plotted in Figure 9 for comparison. Zone A and Zone B from IEC Standard 60034-1 is also plotted.

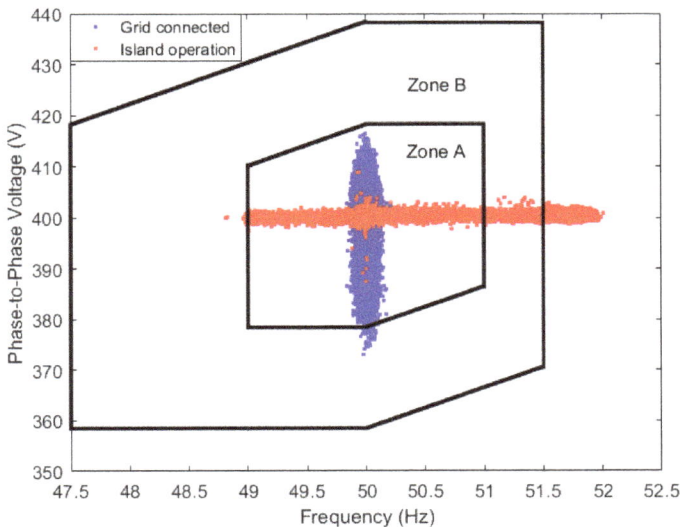

Figure 9. 10 min average values of the phase-to-phase voltage with the corresponding frequency for islanded operation and grid operation with IEC Standard 60034-1 Zone A and Zone B plotted.

The frequency limits, voltage limits and the amount of time in which the nanogrid during islanded operation was within Zone A, Zone B and outside Zone B can be seen in Table 2.

Table 2. Limits for each zone described in IEC Standard 60034-1 and the amount of time the nanogrid operated in each defined zone during islanded operation.

Definition	Within Zone A	Within Zone B	Outside Zone B
Allowed frequency variation	±2%	+3% and −5%	-
Allowed voltage variation	±5%	±10%	-
Amount of 10 min average values within specified zone	79.3%	3.3%	17.4%

The values that are outside Zone A are caused by the curtailment of solar production since the frequency is higher than 51 Hz. The phase-to-phase voltage never exceeds the limits of Zone A. The grid connected 10 min average values are within Zone A for 99.91% of the time and 0.09% in Zone B. It can be seen in Figure 9 that the voltage varies more for the grid connection than islanded operation which also causes some grid connected values to end up in Zone B.

3.1. Single Phase Induction Motors

Single phase induction motors have a start winding that only operates for a few seconds to get the motor spinning. During those few seconds the start winding draws a large current. The timing of the centrifugal switch that disconnects the start winding on some single phase induction motors might get affected with larger frequency variations than what the motor was designed for.

The centrifugal switch disconnects at about 75 to 90% of rated motor speed [16,17]. In a 50 Hz grid that range would correspond to 37.5 to 45 Hz which is a frequency range that can be partly observed in the measurements from the islanded operation.

If a single phase induction motor with a centrifugal switch would start at a supply frequency less than the disconnection speed, the centrifugal switch would not disconnect and leave the start winding operational until the supply frequency increases sufficiently. Such a case could cause the start winding to get damaged or become non-operational. The occasion where the start winding could get damaged is when the nanogrid recovers from an interruption that followed shortly after the frequency reached about 55 Hz. Such a case can be seen in Figure 6 where the frequency was below 45 Hz for about 8 s. A total of 12 occurrences where the frequency stayed below 45 Hz during about 8 s happened during the 48 week measurement time period. In one occurrence the frequency stayed at 43.3 to 44 Hz for about 63 s.

3.2. Computer Power Supplies

Some power supplies for computers follow the ATX12V design specifications which require that the power supply should work between 47 and 63 Hz [6]. Intel power supply design specifications also specify a frequency range of 47 to 63 Hz [7]. Therefore only frequencies below 47 Hz could be a problem since the nanogrid frequency never exceeds 55.2 Hz. Frequencies below 47 Hz occurred 13 times in the 48 week measurements where 12 had duration of about 12 s and one for about 63 s.

3.3. Transformers

IEC Standard 60076-1 [8] states that single phase transformers with larger power rating than 1 kVA and 3-phase transformers with larger power rating than 5 kVA must withstand +5% V/f ratio variation from rated V/f ratio at rated power and frequency. If the voltage would remain constant at rated voltage, the maximum allowed frequency drop would be down to 47.62 Hz for a +5% V/f ratio. The transformer should also withstand a V/f ratio of +10% from rated V/f ratio at no load which correspond to a frequency of 45.5 Hz at rated voltage. The frequency dropped to between 47.6 and 45.5 Hz for 12 times with duration of about 8 to 12 s in the 48 week measurements. For one occasion the frequency dropped below 45.5 Hz for about 63 s. During these instances a transformer might get affected.

3.4. Clocks and Harmonic Filters

Other types of equipment that can be affected by the frequency variations are harmonic filters since they can become de-tuned during periods where there is a large frequency deviation from rated frequency [18]. Clocks that depend on the supply frequency will also be affected. But since over frequency in the nanogrid will be compensated by the AFA, clocks could temporarily get affected during daytime. At 52 Hz the clocks would be off by about two min every ho. In for instance Figure 3, the offset by the evening would be about 14 min.

3.5. Equipment Testing

In order to test the effects on home appliances for the large frequency variations seen in the nanogrid, the test procedure in IEC Standard 61000-4-28 [18] could be used. The frequency test level 2 for equipment for residential customers connected to the low voltage grid is +4% and −6% which for a 50 Hz system corresponds to 47–52 Hz. The transition period from rated frequency to the tested frequency, is 10 s. Since the frequency variations are larger in the nanogrid, test level 4 could be used which applies for non-interconnected networks where misoperation of equipment is critical. Test level 4 uses ±15% (42.5–57.5 Hz) which corresponds closest to the frequency variations measured in the nanogrid. The transition period from rated frequency to the tested frequency is 1 s in test level 4. This is something that corresponds closer to what can sometimes be seen in the nanogrid, see for example Figure 6. Test level 4 is also more appropriate if one was to consider that some home appliances might be critical for maintaining a normal life in the residence. But since the frequency variations in the nanogrid are between 41.3 and 55.2 Hz at 1 s resolution, test level X could be used where the frequency range can be adjusted further. In order to establish the impact on equipment operation for the nanogrid reviewed in this paper, the frequency test level should be at least +10.5% and −17.4% with a transitional period of 1 s.

4. Relationship between Frequency Variations and Interruptions

The interruptions that occurred during the 48 week measurement time period could be divided into three groups according to how the nanogrid transitioned between different operational modes:

Group 1 (Island-interruption-grid): Interruptions during islanded operation that transition into grid operation.
Group 2 (Grid-interruption-island): Interruptions during grid operation that transition into islanded operation.
Group 3 (Island-interruption-island): Interruptions during island operation that transition into islanded operation.

Grid to grid interruptions are not considered since the nanogrid internal energy system is not operational in those cases. The individual downtimes of the interruptions for each three groups are plotted in an empirical CDF in Figure 10 where the longest interruption of 1.97 h has been truncated to give a better visualization of the plot. The primary reason to why the interruptions in each group occurred is unknown.

The total number of interruptions and downtimes for the respective groups can be seen in Table 3. Note that the majority of the interruptions last less than 2 s.

In Group 2, 28 interruptions happened shortly after the frequency in the nanogrid reached about 55 Hz and lasted for about 0.9 s. These are the same type of interruptions that can be seen in Figures 5 and 6 where there is a 10 to 50 second long connection to the utility grid after the frequency reached about 55 Hz. These interruptions amount to 39.4% of the total number of interruptions in 48 weeks and 78% of Group 2. This means that there is a possibility that the surplus of energy in the nanogrid causes approximately 39.4% of the total number of interruptions in the nanogrid. However, the downtime of these interruptions only corresponds to 0.19% of the total downtime in the nanogrid.

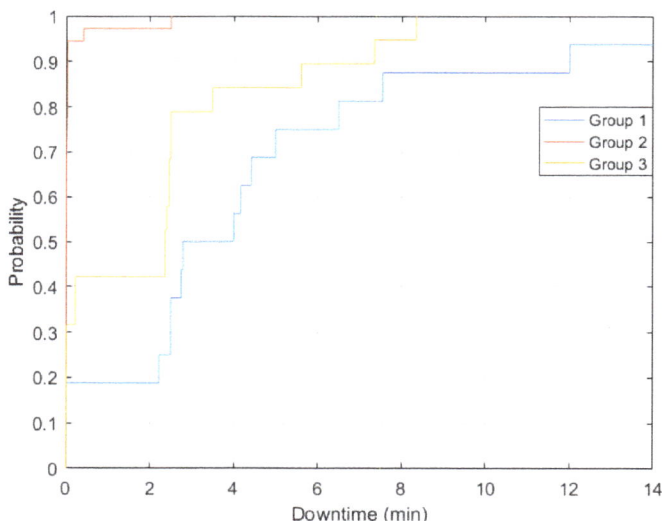

Figure 10. CDF of individual downtimes for the interruptions in each three groups. The plot is truncated at 14 min since there was one interruption in Group 1 that lasted for 1.97 h.

Table 3. Interruptions and downtime for the different groups.

Group	Number of Interruptions	Downtime	Number of Interruptions < 2 s
Group 1	16	2.91 h	3 (19% of group 1)
Group 2	36	3.47 min	33 (92% of group 2)
Group 3	19	42.4 min	6 (32% of group 3)
All groups	71	3.67 h	42 (59% of all groups)

The number of transitions between operational states in Group 1 and Group 2 is approximately 1400 in 48 weeks where around 1000 transitions are caused by the 55 Hz phenomena shown in Figures 5 and 6.

The probability of having an interruption in Group 1 and 2 with regards to the number of transitions between islanded operation and grid operation is 1.1% and 2.5%, respectively, for the 48 weeks.

The probability of having an interruption in Group 2 when the frequency reaches 55 Hz is around2.8% and if the 55 Hz transitions are excluded the probability is around 2%. Since the probability for an interruption is higher when the frequency reaches 55 Hz, the nanogrid could be more sensitive to interruptions when the nanogrid transitions to grid operation.

5. Possible Solutions for Reducing over Frequency in Islanded Operation

The over-frequency in the nanogrid is caused by the FSPC used by the Sunny Island battery inverters to signal the solar inverters to regulate the power production. This is done when there is not enough load connected in the nanogrid during islanded operation. The over-frequency caused by the FSPC will activate the AFA which lowers the frequency below 50 Hz to compensate for the occurred over frequency. If instead a direct link with a cable between the battery inverter and solar inverter would be used, the signaling with the power frequency could be avoided and therefore possibly eliminate the frequency variations. However, it is unclear if the lowest frequency variations that happen when the islanded operation initiates after an interruption would be eliminated with this method.

If this solution can't be done practically, a simple solution would be to have a large resistive load (dump load) that can be activated when the frequency starts to rise above 51 Hz. The solution of using dump loads to regulate the power frequency in islanded operated microgrids when there is an excess amount of power in the system is described in for instance [19,20].

Another solution would be to increase the storage capacity (which is under construction) and/or increase the electrolyzer power in order to create more hydrogen when there is not enough consumption.

If an increase in energy storage is not feasible and if the objective is to reduce the loss of potential power production, one could shift some of the loads towards the day when the solar power production is occurring. In a single house nanogrid, such loads could be for example the dishwasher, washing machine, electric vehicles, air conditioning units or heat pumps. If such an approach would be taken, the service life of the battery would also increase since the cycling of the battery during the night is reduced.

These solutions could reduce the large frequency variations that go beyond the limits in product and grid standards described in this article. An increase in energy storage and consumption when the power production occurs would also be necessary in order to reduce the amount of transitions between islanded operation and grid operation. That could in turn reduce the amount of interruptions that occur during such transitions.

6. Conclusions

The 10 s average frequency variations in the nanogrid during islanded operation are outside the range set by EN 50160 for systems without synchronous connection to an interconnected system for 89.6% of the 48 weeks. However, for Standard EN 50160/A1 which applies in Norway the frequency variations are outside the limits for 25% of 48 weeks.

The lower and upper allowed 10 s average frequency limit (52.5 to 57.5 Hz) defined by EN 50160 is not surpassed but for EN 50160/A1 the maximum allowed range of 49 to 51 Hz is surpassed.

The frequency variations between 51 and 52 Hz are caused by the FSPC used by the Sunny Island inverters to curtail production from the solar PV installation when there is not enough consumption. The larger frequency variations from 52 to 55 Hz occur when there is not enough consumption during the daytime and when the FSPC increases the frequency towards 55 Hz to shut down the solar inverters in order to synchronize with the utility grid.

The lowest frequency values of about 41 to about 49 Hz are caused by short interruptions after the frequency reached about 55 Hz. The frequency values at about 49 Hz are caused by the AFA compensating for the occurred over frequency in order to enable clocks to run at the correct time.

There might be some adverse effects on certain equipment of these frequency variations. For instance, AC motors might be affected since 17.4% of the total time in islanded operation AC motors will operate outside the limits described by IEC Standard 60034-1. Single phase induction motors might be affected if they are started just after the short interruptions that can occur when the frequency has reached 55 Hz. This is since the frequency can be lower than the centrifugal switch opening frequency for about 8 to 63 s at 13 occurrences which in turn could cause damage to the start winding. Computer power supplies and transformers could also be affected for 13 times in the 48 week measurements for duration of about 8 to 63 s at each occurrence. The frequency variations that go beyond the allowed range described in grid and product standards could be eliminated by increasing the consumption by for instance shifting consumption to the daytime when the production occurs. Another solution would be to increase the energy storage in order to store the excess generated power or have a direct link between the Sunny Island battery inverter and solar inverter to avoid the communication through the power frequency.

Approximately 39.4% of the total number of interruptions could also possibly be eliminated by ensuring that the load is matched to the solar production. Since transitions between islanded operation and grid operation increase the risk of interruptions, a constantly islanded nanogrid could have fewer interruptions than what this case study has presented. It is unclear if the reliability of the nanogrid would increase with the removal of the possibility of connecting to the grid since there could be instability in the system causing the interruptions in Group 2 presented in this paper.

Author Contributions: Conceptualization, J.N., S.K.R. and M.H.J.B.; Methodology, J.N.; Software, J.N.; Validation, J.N., S.K.R. and M.H.J.B.; Formal Analysis, J.N.; Investigation, J.N.; Resources, S.K.R. and M.H.J.B.; Data Curation, J.N.; Writing-Original Draft Preparation, J.N.; Writing-Review & Editing, J.N., S.K.R. and M.H.J.B.; Visualization, J.N.; Supervision, S.K.R. and M.H.J.B.; Project Administration, S.K.R. and M.H.J.B.; Funding Acquisition, S.K.R. and M.H.J.B.

Funding: This paper has been funded by Skellefteå Kraft Elnät and Rönnbäret foundation.

Conflicts of Interest: The authors declare no conflict of interest.

References

1. Hatziargyriou, N.; Asano, H.; Iravani, R.; Marnay, C. Microgrids. *IEEE Power Energy Mag.* **2007**, *5*, 78–94. [CrossRef]
2. Burmester, D.; Rayudu, R.; Seah, W.; Akinyele, D. A Review of Nanogrid Topologies and Technologies. *Renew. Sustain. Energy Rev.* **2017**, *67*, 760–775. [CrossRef]
3. *Cenelec Standard EN 50160, Voltage Characteristics of Electricity Supplied by Public Electricity Networks*; European Committee for Electrotechnical Standardization: Brussels, Belgium, 2010.
4. *Cenelec Standard EN 50160/A1, Voltage Characteristics of Electricity Supplied by Public Electricity Networks*; European Committee for Electrotechnical Standardization: Brussels, Belgium, 2015.
5. *IEC Standard 60034-1, Rotating electrical machines—Part 1: Rating and performance*; International Electrotechnical Commission: Geneva, Switzerland, 2017.
6. Intel Corporation. *ATX12V, Power Supply Design Guide*, version 2.2; Intel Corporation: Santa Clara, CA, USA, 2005.
7. Intel Corporation. *Design Guide for Desktop Platform Form Factors*, revision 1.31; Intel Corporation: Santa Clara, CA, USA, 2013.
8. *IEC Standard 60076-1:2011, Power transformers-Part 1: General*; International Electrotechnical Commission: Geneva, Switzerland, 2011.
9. Rönnberg, S.K.; Bollen, M.H.J.; Nömm, J. Power Quality Measurements in a Single House Microgrid. In Proceedings of the CIRED 24th International Conference on Electricity Distribution, Glasgow, Scotland, 12–15 June 2017; pp. 818–822.
10. SMA. *PV Inverters, Use and Settings of PV Inverters in Off-Grid Systems*, version 4.2; SMA: Niestetal, Germany, 2014.
11. SMA. *Sunny Island 3324/4248 Installation Guide*, version 4.0; SMA: Niestetal, Germany, 2005.
12. *IEEE Standard 1789-2015, IEEE Recommended Practices for Modulating Current in High-Brightness LEDs for Mitigating Health Risks to Viewers*; The Institute of Electrical and Electronics Engineers: New York, NY, USA, 2015.
13. Gil-de-Castro, A.; Rönnberg, S.K.; Bollen, M.H.J. Light intensity variation (flicker) and harmonic emission related to LED lamps. *Electr. Power Syst. Res.* **2017**, *146*, 107–114. [CrossRef]
14. Bollen, M.H.J.; Gu, I.Y.H. *Signal Processing of Power Quality Disturbances*, 1st ed.; Wiley-IEEE Press: Hoboken, NJ, USA, 2006; p. 159.
15. Rajput, R.K. *Alternating Current Machines*, 1st ed.; Firewall Media: New Delhi, India, 2002; p. 435.
16. Brumbach, M.E. *Industrial Electricity*, 9th ed.; Cengage learning: Boston, MA, USA, 2017; p. 385.
17. Shultz, G.P. *Transformers and Motors*, 1st ed.; Elsivier: New York, NY, USA, 1989; p. 129.
18. *IEC Standard 61000-4-28, Electromagnetic Compatibility (EMC)–Part 4–28: Testing and Measurement Techniques-Variation of Power Frequency, Immunity Test for Equipment with input Current not Exceeding 16 A per phase*, edition 1.2; International Electrotechnical Commission: Geneva, Switzerland, 2009.
19. Serban, E.; Serban, H. A Control Strategy for a Distributed Power Generation Microgrid Application with Voltage and Current Controlled Source Converter. *IEEE Trans. Power. Electron.* **2010**, *25*, 2981–2992. [CrossRef]
20. Baudoin, S.; Vechiu, I. Review of Voltage and Frequency Control Strategies for Islanded Microgrid. In Proceedings of the 16th International Conference on System Theory, Control and Computing (ICSTCC), Sinaia, Romania, 2–14 October 2012.

![energies logo] *energies*

MDPI

Article

An Analysis of Voltage Quality in a Nanogrid during Islanded Operation

Jakob Nömm *, Sarah K. Rönnberg * and Math H. J. Bollen *

Electric Power Engineering, Luleå University of Technology, 931 87 Skellefteå, Sweden
* Correspondence: jakob.nomm@ltu.se (J.N.); sarah.ronnberg@ltu.se (S.K.R.); math.bollen@ltu.se (M.H.J.B.)

Received: 21 January 2019; Accepted: 13 February 2019; Published: 15 February 2019

Abstract: Voltage quality data has been collected in a single house nanogrid during 48 weeks of islanded operation and 54 weeks of grid-connected operation. The voltage quality data contains the voltage total harmonic distortion (THD), odd harmonics 3 to 11 and 15, even harmonics 4 to 8, voltage unbalance, short-term flicker severity (Pst) and long-term flicker severity (Plt) values, and voltage variations at timescales below 10 min. A comparison between islanded and grid-connected operation values was made, were some of the parameters were compared to relevant grid standard limits. It is shown that some parameters exceed the defined limits in the grid-standards during islanded operation. It was also found that the islanded operation has two modes of operation, one in which higher values of the short circuit impedance, individual harmonic impedance, harmonic voltage distortion and voltage unbalance were reached.

Keywords: harmonics; islanded operation; nanogrids; power quality; voltage unbalance

1. Introduction

Microgrids and nanogrids can provide economical gains in the form of price reductions for consumers and increased revenue for grid owners [1]. They could also provide an improved technical solutions such as energy loss reduction and better reliability than a regular utility connection for certain geographical areas [1,2]. The international council on large electric systems (CIGRE) WG C6.22 defines microgrids as: "electricity distribution systems containing loads and distributed energy resources, (such as distributed generators, storage devices, or controllable loads) that can be operated in a controlled, coordinated way either while connected to the main power network or while islanded" [3]. The term nanogrid was suggested in [4] for defining a small microgrid, which could be a single residential house.

There is a lack of published papers that contain voltage quality measurements that span several months or years for nanogrids in islanded operation. These measurements are needed to establish the differences in performance between islanded operation and grid-connected operation. The measurements would also make it possible to evaluate if problems can appear for connected equipment during islanded operation.

In this paper, long term measurements of voltage quality are presented that have been collected in a single house nanogrid during 48 weeks of islanded operation and 54 weeks of grid-connected operation. Some of the measured voltage quality parameters have been compared with the limits defined in standards EN 50160 [5] and IEEE 519-2014 [6]. The specified standards do not include voltage quality limits for islanded operation, so the limits in the standards are only used as a reference for islanded operation.

The main contribution of this paper is the analysis and presentation of long-term voltage quality measurements collected in a nanogrid during islanded operation. All the used equipment in the nanogrid is commercially available and therefore similar performance is expected for other nanogrids like the one presented in this paper.

The Nanogrid

The single house nanogrid that is studied in this paper is located in the southern part of Sweden and has a 22.6 kWp solar installation, 144 kWh lead acid battery storage, 1100 kWh hydrogen storage and a 15 kVA diesel backup generator. The nanogrid is designed to operate as a 50 Hz three-phase system where each phase has a phase-to-neutral voltage of 230 V root mean square (RMS). The consumption in the nanogrid consists of ordinary household appliances, two electric cars, a three-phase heat pump and an electrolyzer for the production of hydrogen. A simplified schematic of the nanogrid energy system can be seen in Figure 1.

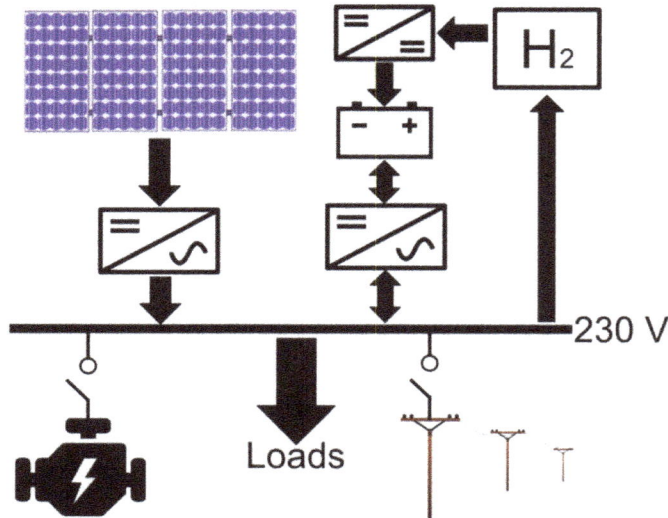

Figure 1. Simplified schematic of the energy system for the nanogrid. Reproduced with permission from [7], Nömm, J.; Rönnberg, S.K.; Bollen, M.H.J. An Analysis of Frequency Variations and its Implications on Connected Equipment for a Nanogrid during Islanded Operation. Energies 2018, 11, 2456.

The nanogrid is designed to run primarily on the produced solar power and the stored energy in the batteries and hydrogen tanks. During the day, the solar panels will supply the energy to the loads where the excess solar power will charge the batteries and power a 5 kW electrolyzer to convert electricity to hydrogen that is stored in high-pressure tanks. During the night, the batteries are the main supply of energy to the nanogrid; if the battery charge drops below 30%, a 5 kW fuel cell will convert hydrogen to electricity to charge the battery. If there is a malfunction in the primary energy system, the nanogrid will connect to the low-voltage utility grid. The nanogrid also connects to the utility grid if the stored energy in the batteries and hydrogen tanks is depleted. The backup diesel generator in the nanogrid is designed to start only if both the low voltage utility grid and the primary energy system in the nanogrid fail to operate. The backup diesel generator operated for 43 h during the 48 week islanded operation measurement period.

The nanogrid switched to grid-connected operation mainly due to lack of energy stored in the batteries and hydrogen tanks. To avoid this, additional hydrogen storage is under construction. For more information regarding the nanogrid see [7] and [8].

2. Methodology

The measurements have been collected by an Elspec G4430 (Elspec, Caesarea, Israel) connected at the load-output of the SMA Multicluster Box (SMA, Niestetal, Germany). The SMA Multicluster Box is used in the nanogrid since there are three independent solar installations, two on the roof, each with a capacity of 10 kW and one on the facade with 2.6 kW power rating.

The parameters measured for the comparison to the limits described in EN 50160 and IEEE 519-2014 are the voltage Total Harmonic Distortion (THD), Individual odd harmonics 3rd to 11th and 15th, even harmonics 4th to 8th, Pst, Plt, and voltage unbalance. Two parameters that were also measured with no relation to any standard were the very short variations (VSV) of the voltage and RMS value of the neutral current. For the measured parameters, the total time in islanded operation was 48 weeks and 54 weeks in grid-connected operation. The 48 and 54 week measurements are assembled from shorter time windows in which the nanogrid was in islanded or grid-connected operation.

For the voltage THD measurements, there was a measurement period of 29 weeks of the total 48 weeks where the nanogrid operated continuously in islanded operation. These measurements were used to see the daily voltage THD variations.

For the analysis of the individual harmonics, the odd harmonics 3rd to 11th and 15th and even harmonics 4th to 8th were chosen since they all surpass the limits defined in either standard EN 50160 or IEEE 519-2014 sometime during the 48 week islanded operation measurements.

Another measurement period of about 8 weeks in islanded operation and about 5 weeks in grid-connected operation was used to study the variations with time in the short circuit impedance measured as the voltage drop against a current rise of larger than 4 A within two cycles.

3. Results

3.1. Total harmonic Distortion

The cumulative distribution function (CDF) for the 10 min values of the voltage THD during 48 weeks in islanded operation and 54 weeks in grid-connected operation can be seen in the upper part of Figure 2.

In the lower part of Figure 2, the corresponding CDF for the 3 s values is plotted. As expected, the 3 s voltage THD values reach higher values than the 10 min values during both islanded and grid-connected operation. It can also be observed in Figure 2 that the voltage THD is always higher during islanded operation than during grid-connected operation. The maximum values and total average values are also higher for islanded operation which can be seen in Table 1.

Table 1. 95% Confidence interval (CI) for the 10 min values, the maximum 10 min value and total average value for all three phases in islanded and grid-connected operation.

Operational State	95% CI 10 min Value	Max 10 min Value	95% CI 3 s Value	Max 3 s Value	Total Average Value
Islanded	1.34 to 8%	7.83 to 13.01%	1.28 to 8.06%	18.3 to 21.9%	2.23 to 3.82%
Grid-connected	0.77 to 1.88%	2.41 to 2.47%	0.76 to 1.88%	$\approx$2.5%	1.08 to 1.44%

In Figure 3, the average voltage THD variations for each hour of the day for 29 weeks can be seen for all three phases. During 29 out of 48 weeks the nanogrid operated in islanded operation that lasted continually throughout the day without connections to the utility grid.

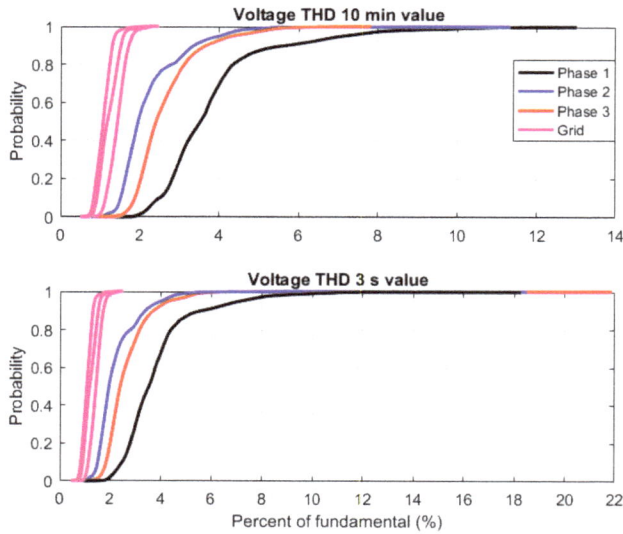

Figure 2. Cumulative distribution function (CDF) for the 10 min voltage Total Harmonic Distortion (THD) for the nanogrid during islanded and grid-connected operation (**top**) and the CDF for the 3 s voltage THD during grid-connected and islanded operation (**bottom**). Note that the horizontal axis is different for each plot.

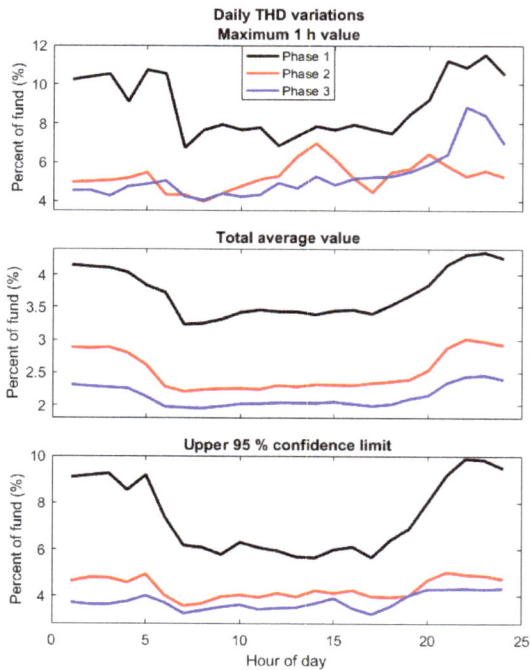

Figure 3. The 1 h maximum THD value, total average THD value and upper 95% confidence limit for every hour during the day for 29 weeks of continuous islanded operation. The black, red and blue color represents the three phases. Note the difference in vertical scale for each plot.

It can be seen that the maximum 1 h values are reached during the night for phase 1 and 3. For phase 2, the maximum occurs in the middle of the day. The total average value and 95% confidence limit for each hour during the day reach the highest values in the morning and night for all 3 phases. Phase 3 has however a smaller variation between night and day in the total average value and 95% confidence limit as phase 1 and 2. A more detailed view of the voltage THD variations during one day can be seen in [9].

The trend in Figure 3 shows larger THD values during the night than during the day. This indicates that when the solar production starts, the voltage THD level drops due to more parallel sources being activated and increases when there are fewer parallel sources available. One example when the voltage THD suddenly increases during the evening at about the time when the sun sets can be seen in Figure 4. It can be seen that even though the active power remains almost constant the voltage THD increases for all three phases which also can be seen in the voltage waveform. The current THD also increases for phase 1 and 3 and decreases for phase 2 which can also be seen in the current waveform. The reactive power changes somewhat for the three phases and the frequency throughout Figure 4 was around 49 Hz.

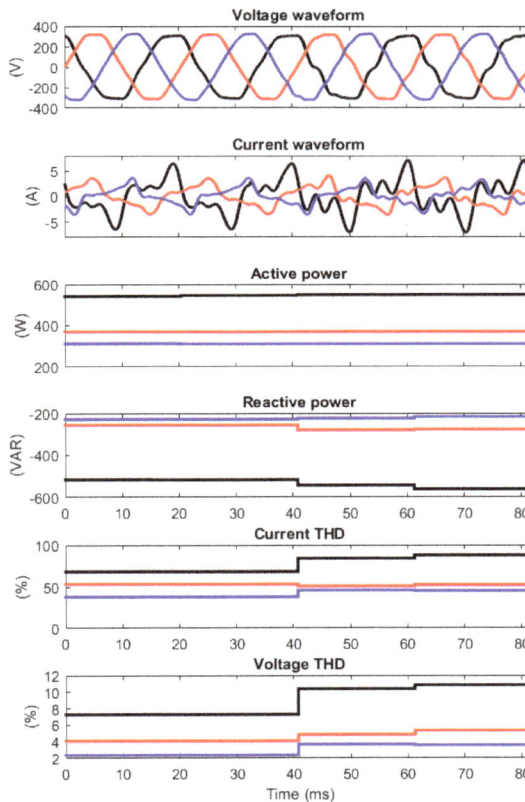

Figure 4. Example of an occasion when the voltage THD increases in the evening. The voltage waveform is at the top followed by the current waveform, active power, reactive power, current THD and voltage THD. The black, red and blue color represents the three phases.

3.2. Individual Harmonics

In Figure 5, the odd voltage and current harmonics 3 to 9 are presented for the 1-cycle period ending at about 20 ms and 80 ms in Figure 4. The 3rd voltage harmonic is about the same for phase 1 and 2 and slightly lower for phase 3. After the transition from 8 to 12% voltage THD (seen at around 40 ms in Figure 4), the 5th voltage harmonic increases for two of the three phases and the 7th and 9th voltage harmonic also increase for all three phases. All of the current harmonics increase for phase 1 but the 3rd harmonic for phase 2 decreases by 1.4% and the 3rd and 7th harmonic for phase 3 decrease by 4.35 and 1.34% respectively.

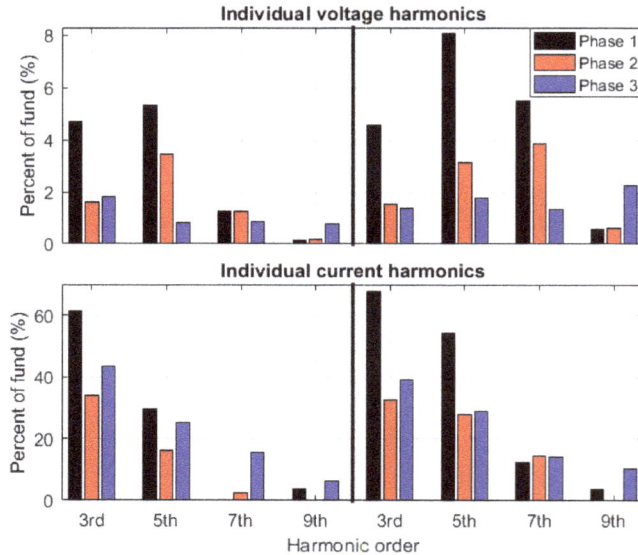

Figure 5. Individual harmonics of the voltage THD in Figure 4 where the left side is the individual harmonics at the end of the 1-cycle period at about 20 ms and the right side at about 80 ms.

The 95% confidence limit for the 10 min values for the odd voltage harmonics 3rd to 11th and 15th and even voltage harmonics 4th to 8th can be seen in the upper part of Figure 6.

The maximum 10 min value can be seen in the lower part of Figure 6. It can be seen that both the maximum 10 min value and the 95% confidence limit value are higher during islanded operation. The odd harmonics differ the most in magnitude from the grid-connected measurements, except the 95% confidence limit value for the 15th harmonic. Phase 1 has higher 95% confidence limits for the odd harmonics until the 11th. In the maximum 10 min values, the even harmonics are several times higher in islanded operation compared to grid-connected operation. The 95% confidence limit for the even harmonics is close to zero for the grid-connected operation.

At a shorter time scale of 3 s, which can be seen in Figure 7, the 95% confidence limit is about the same as the 95% 10 min values in Figure 6. But the maximum 3 s value is larger than the 10 min maximum value. The even harmonics have the highest increase from 10 min maximum values to 3 s maximum values. Phase 2 had the lowest 10 min maximum value for most of the harmonic orders, but for the 3 s values it has the highest value for most of the harmonic orders.

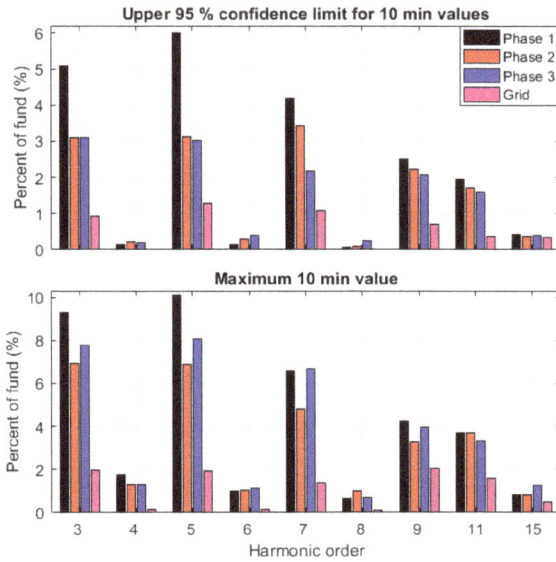

Figure 6. 10 min 95% confidence limit value (**top**) and the maximum 10 min value in the measurements (**bottom**) for the odd harmonics 3rd to 11th and 15th and even harmonics 4th to 8th. The black, red and blue color represents phase 1 to 3 in islanded mode. Purple color represents all three phases during grid-connected operation since the distortion differs substantially less between the phases in comparison to islanded operation.

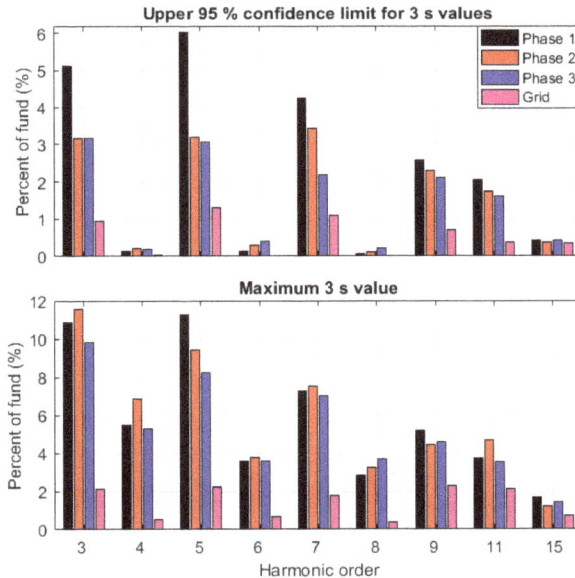

Figure 7. 3 s maximum 95% confidence limit value (**top**) and the maximum 3 s value in the measurements (**bottom**) for the odd harmonics 3rd to 11th and 15th and even harmonics 4th to 8th. The black, red and blue color represents phase 1 to 3 in islanded mode. Purple color represents all three phases during grid-connected operation.

The total average values for both grid-connected and islanded operation can be seen in Table 2.

Table 2. Total average value for each measured voltage harmonic for all three phases.

Operational State	3rd (%)	4th (%)	5th (%)	6th (%)	7th (%)	8th (%)	9th (%)	11th (%)	15th (%)
Islanded	0.95 to 1.96	0.01 to 0.02	0.83 to 2.0	0.016 to 0.04	0.61 to 1.45	0.008 to 0.016	1.20 to 1.38	0.63 to 0.66	0.20 to 0.18
Grid-connected	0.30 to 0.52	≈ 0.001	0.66 to 0.90	$\approx 10^{-4}$	0.55 to 0.60	$\approx 10^{-5}$	0.34 to 0.44	0.14 to 0.17	0.10 to 0.15

3.3. Voltage Unbalance

The 10 min voltage unbalance for both islanded operation and grid-connected operation can be seen in Figure 8. The voltage unbalance is for the majority of the measured time lower in islanded operation than in grid-connected operation. The maximum 10 min voltage unbalance was 4.6% for islanded operation and 1.77 for grid-connected operation. In Figures 14–16 the occurrence of the maximum voltage unbalance value can be seen. A summary of the 95% CI, maximum 10 min value and total average value in islanded and grid-connected operation can be seen in Table 3.

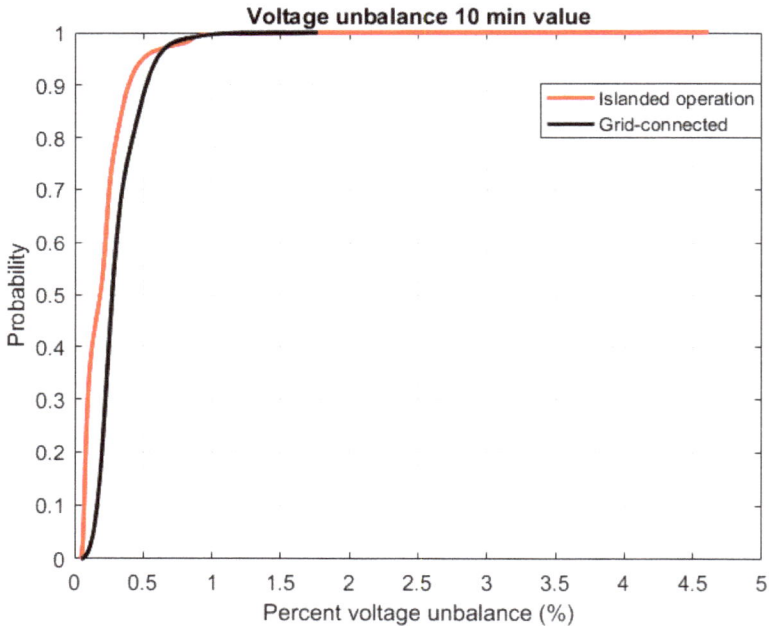

Figure 8. CDF for the 10 min voltage unbalance values in the nanogrid during islanded and grid-connected operation.

Table 3. 95% confidence interval, the maximum 10 min value and total average value for all three phases in islanded and grid-connected operation.

Operational State	95% CI 10 min Value	Max 10 min Value	Total Average Value
Islanded	0.06 to 0.72%	4.6%	0.22%
Grid-connected	0.11 to 0.68%	1.77%	0.31%

3.4. Voltage Fluctuations

The CDF of the Pst values for grid-connected and islanded operation can be seen in Figure 9. The islanded operation data reaches lower Pst values than in grid-connected operation for 95% and 33% of the total measured time for phase 2 and 1. Phase 3 has larger Pst values for the majority of the time when comparing to the grid-connected operation Pst values. All three phases have higher maximum values than the grid-connected measurements. A summary of the 95% CI, maximum value and total average value for all three phases in islanded and grid-connected operation can be seen in Table 4.

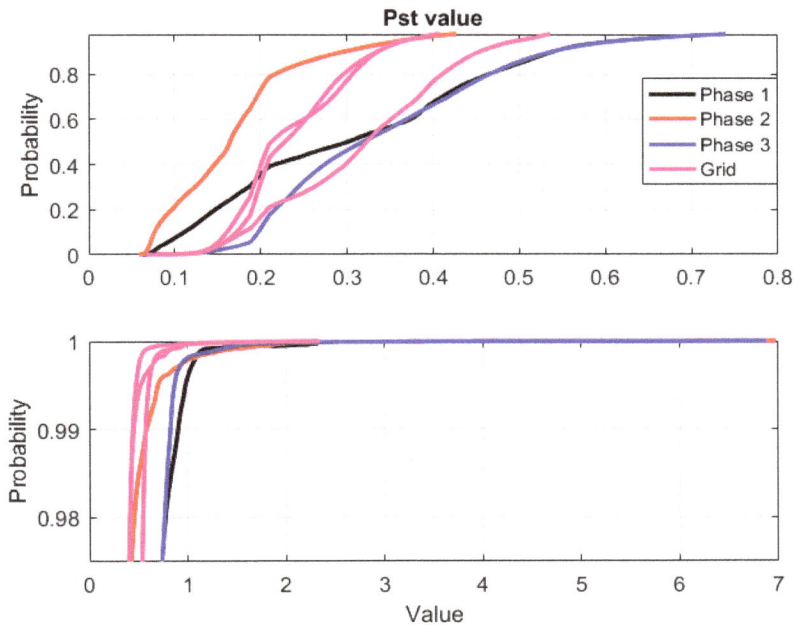

Figure 9. CDF for the Pst values for grid-connected and islanded operation where the interval 0 to 97.5% is displayed at the top and the remaining upper 2.5% is displayed in the bottom. The black, red and blue color represents phase 1 to 3 in islanded mode. Purple color represents all three phases during grid-connected operation.

Table 4. 95% confidence interval, the maximum 10 min value and total average value for all three phases in islanded and grid-connected operation.

Operational State	95% CI	Max Value	Total Average Value
Islanded	0.06 to 0.74	6.43 to 7	0.18 to 0.35
Grid-connected	0.14 to 0.52	2.3 to 2.34	0.24 to 0.32

The CDF of the Plt values for islanded and grid-connected operation can be seen in Figure 10. It can be seen that phase 1 and 2 have lower Plt values for 91% and 35% of the total measured time in islanded operation when compared to grid-connected operation. Phase 3 has lower Plt values for about 45% of the total measured time than the phase with highest Plt values in grid-connected operation. All three phases reach higher maximum Plt values in islanded operation. The 95% CI, maximum value and total average value can be seen in Table 5.

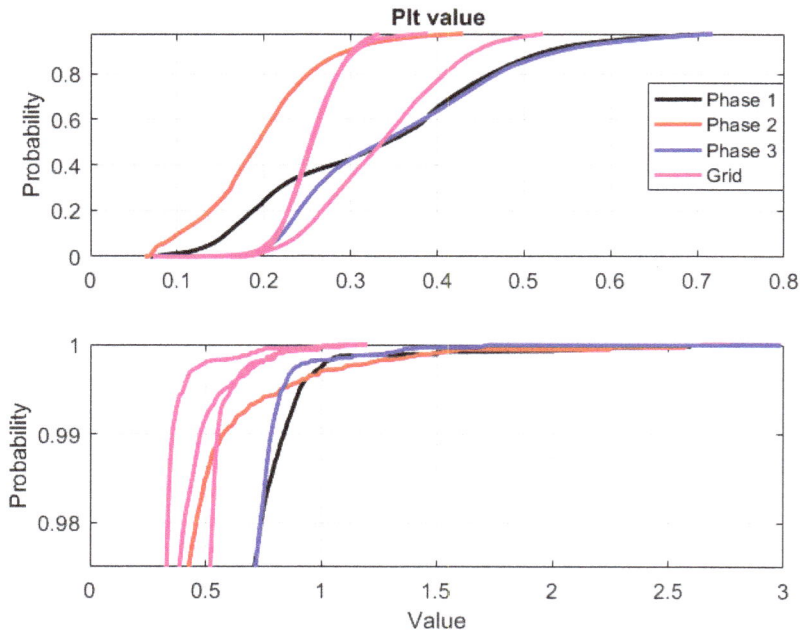

Figure 10. CDF for the Plt values in grid-connected and islanded operation where the interval 0 to 97.5% is displayed at the top and the remaining upper 2.5% is displayed in the bottom. The black, red and blue color represents phase 1 to 3 in islanded mode. Purple color represents all three phases during grid-connected operation.

Table 5. 95% confidence interval, the maximum 10 min value and total average value for all three phases in islanded and grid-connected operation.

Operational State	95% CI	Maximum Value	Total Average Value
Islanded	0.07 to 0.72	2.78 to 2.99	0.21 to 0.36
Grid-connected	0.19 to 0.52	1.16 to 1.2	0.26% 0.34

Voltage Variations below 10 Min Values

The variations in the voltage that are on a shorter time scale than 10 min are calculated using the very short variations (VSV) of the voltage. It was introduced by [10,11] to quantify the difference between very-short time scale (one to several seconds) voltage RMS values and the short time scale (10 min) voltage RMS value. The 10 min and 3 s VSV values for grid-connected and islanded operation can be seen in Figure 11 where both the 10 min and 3 s VSV values are lower for the majority of the time for islanded operation. The 3 s VSV values reach higher maximum values in islanded operation. A summary of the 95% CI, maximum values and total average value can be seen in Tables 6 and 7.

Table 6. 95% confidence interval, the maximum 10 min value and total average value for all three phases in islanded and grid-connected operation.

Operational State	95% CI 10 min	Max 10 min Value	Total Average Value
Islanded	0.013 to 0.56	4.6 to 6.34	0.095 to 0.15
Grid-connected	0.18 to 1.9	5.75 to 7.89	0.61 to 0.81

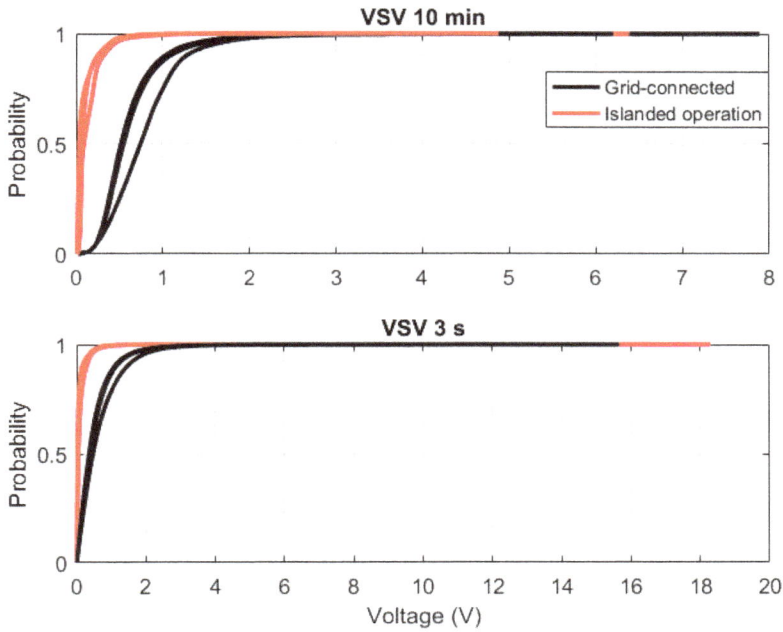

Figure 11. CDF for the 10 min very short variations (VSV) values (**top**) and the 3 s VSV values (**bottom**) for islanded and grid-connected operation.

Table 7. 95% confidence interval, the maximum 3 s value and total average value for all three phases in islanded and grid-connected operation.

Operational State	95% CI 3 s	Max 3 s Value	Total Average Value
Islanded	≈0 to 0.58	15.52 to 18.3	0.07 to 0.11
Grid-connected	0.015 to 2.32	10.6 to 15.7	0.50 to 0.66

3.5. Neutral Current

The CDF for the 10 min RMS neutral current for 48 weeks of islanded operation and 54 weeks of grid-connected operation can be seen in Figure 12.

The 50 Hz component was always several times larger than the zero sequence components where the 3rd harmonic reached a maximum value of 6 A during the 48 week measurement period in islanded operation. For both islanded and grid-connected operation, the line to neutral voltage was close to 0 V. The neutral current reached about the same maximum values of around 42 A in islanded and grid-connected operation. However, during islanded operation the 10 min values are for the most part higher than the grid-connected values. A summary of the 95% CI, maximum values and total average value can be seen in Table 8.

Table 8. 95% confidence interval, the maximum 10 min value and total average value for all three phases in islanded and grid-connected operation.

Operational State	95% CI 10 min	Max 10 min Value	Total Average Value
Islanded	2.15 to 17.6	41.4	5
Grid-connected	1.2 to 15.1	41.8	3.6

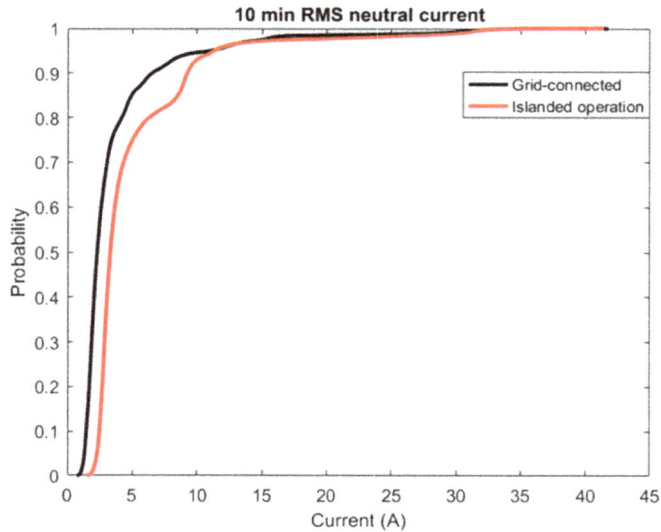

Figure 12. CDF for the 10 min RMS neutral current during islanded and grid-connected operation.

3.6. Operational Modes during Islanded Operation.

The short circuit impedance is extracted from a 13 weeks continuous period. During this 13 week period, the nanogrid operated part of the time in islanded operation and part of the time in grid-connected operation, which can be seen in Figures 13–16. The short circuit impedance was measured as the voltage drop against a current rise of larger than 4 A within two cycles. 4 A was chosen since values lower than 2 A could sometimes not affect the voltage, therefore a larger value of 4 A giving a 2 A margin was chosen. No correlation between the current rise magnitude (4 to 26 A in this paper) and the short circuit impedance were found and no correlation between the starting value of the current (0 to 36 A) and the short circuit impedance was found.

In Figure 13 two different modes of islanded operation can be distinguished, one where the short circuit impedance fluctuates between 0.1 and 1 Ω and one in which it fluctuates between 0.1 and 2 Ω and sometimes reaching values close to 6 Ω but only for phase 2. During the grid-connected operation the values fluctuate between 0.2 and 0.4 Ω. The number of samples for the different phases was 31,000, 730 and 11,000 for phase 1 to 3 respectively, giving rise to differences in the horizontal axis in Figure 13. In Figures 14–16, the 10 min voltage THD, current THD, 3rd, 5th, 7th, 9th harmonic voltage and voltage unbalance are plotted for the 13 week period. The two different modes of islanded operation are again indicated. The voltage unbalance reaches the highest levels of about 4.6% during mode 2 of the islanded operation. The voltage THD, 3rd, 5th and 7th harmonic voltage is also larger in mode 2 for all three phases during the majority of the measured period. The 9th harmonic is for the most part lower in mode 2 for all three phases.

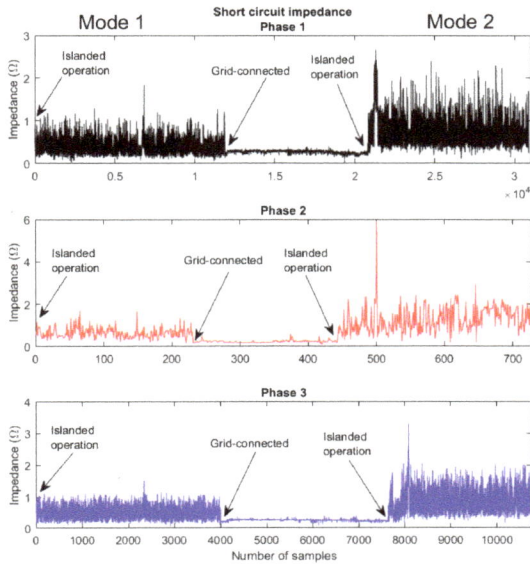

Figure 13. Short circuit impedance for 13 weeks of islanded and grid-connected measurements. The number of samples are around 31,000, 730 and 11,000 for phase 1 to 3 respectively. Note the difference in horizontal scale for each sub-plot.

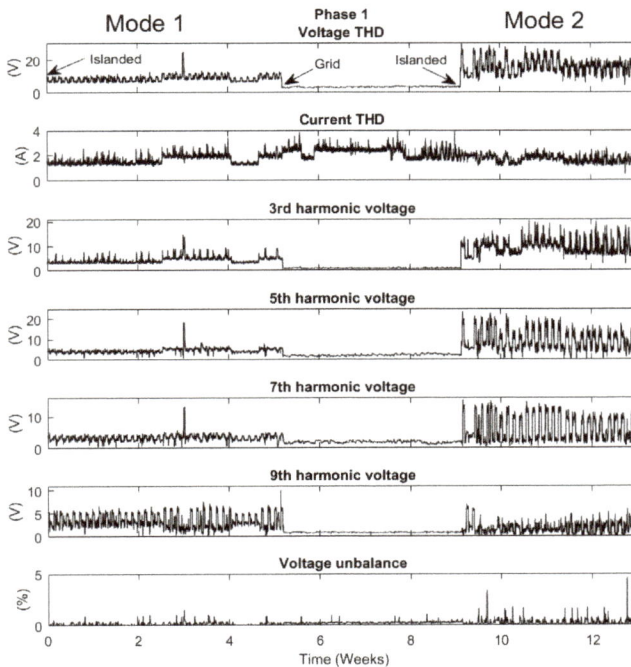

Figure 14. The voltage and current THD, 3rd, 5th, 7th, 9th harmonic voltage and voltage unbalance for phase 1 during a 13 week period in which the samples was extracted for a current rise larger than 4 A within 2 cycles.

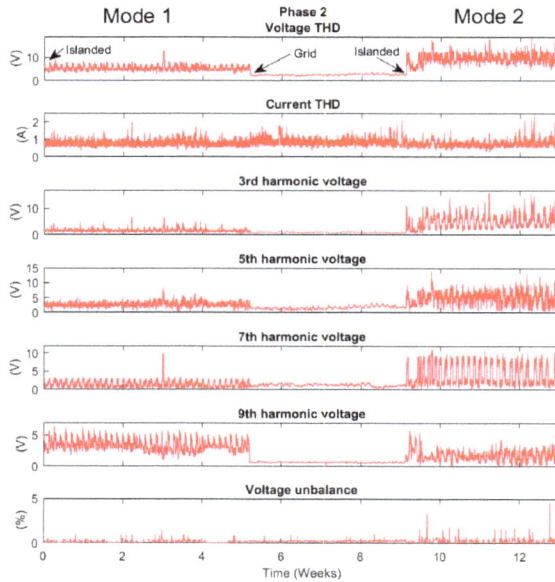

Figure 15. The voltage and current THD, 3rd, 5th, 7th, 9th harmonic voltage and voltage unbalance for phase 2 during a 13 week period in which the samples was extracted for a current rise larger than 4 A within 2 cycles.

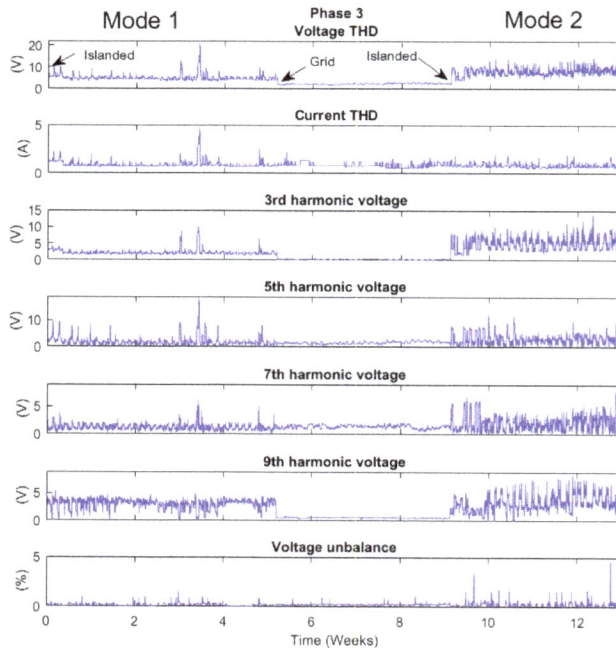

Figure 16. The voltage and current THD, 3rd, 5th, 7th, 9th harmonic voltage and voltage unbalance for phase 3 during a 13 week period in which the samples was extracted for a current rise larger than 4 A within 2 cycles.

The average short circuit impedance for each hour during the day for around 8 weeks in islanded operation seen in Figure 13 can be seen in Figure 17. The blue and red line represents mode 1 and mode 2 respectively seen in Figure 13. The amount of samples for phase 1 to 3 is around 31,000, 730, 11,000 respectively, therefore phase 2 doesn't have samples for some hours during the day. It can be seen that the short circuit impedance is larger during the beginning of the day for phase 1 and larger during the middle of the day for phase 3. For phase 2 there are not enough samples to clearly see the variation during the day. However, for all three phases it can be seen that the short circuit impedance has a higher average for each hour during the day during mode 2 in islanded operation.

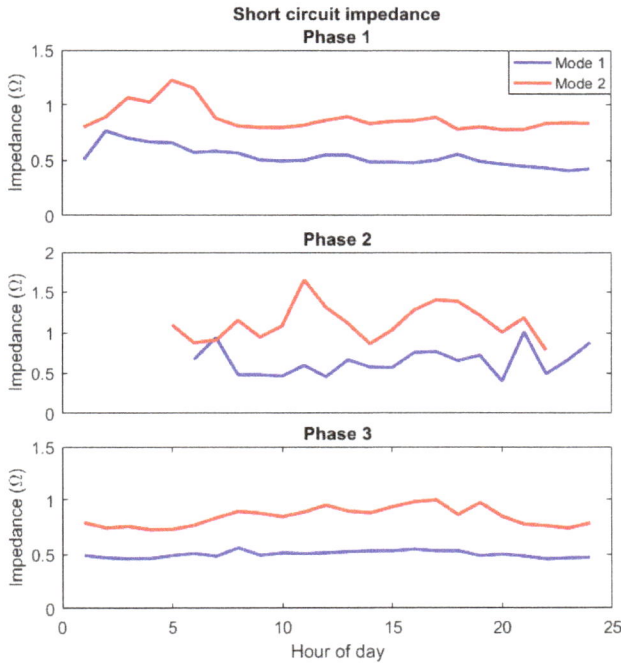

Figure 17. The average short circuit impedance for each hour of the day for around 8 weeks in islanded operation for Mode 1 and Mode 2 seen in Figure 13.

The average values for each hour of the day for around 8 weeks in islanded operation for the 3rd to 9th odd harmonic voltages, currents and voltage THD for mode 1 in Figures 14–16 can be seen in Figure 18. Mode 1 has larger 7th and 9th harmonic voltages and currents during the night for all three phases and the 5th harmonic has larger values for phase 1 and 3. The 3rd harmonic is more evenly distributed across the day where the 3rd harmonic has larger average voltage and current values during the middle of the day. The voltage THD has the largest values during the night which corresponds with the result from the longer 29 week measurements presented in Figure 3.

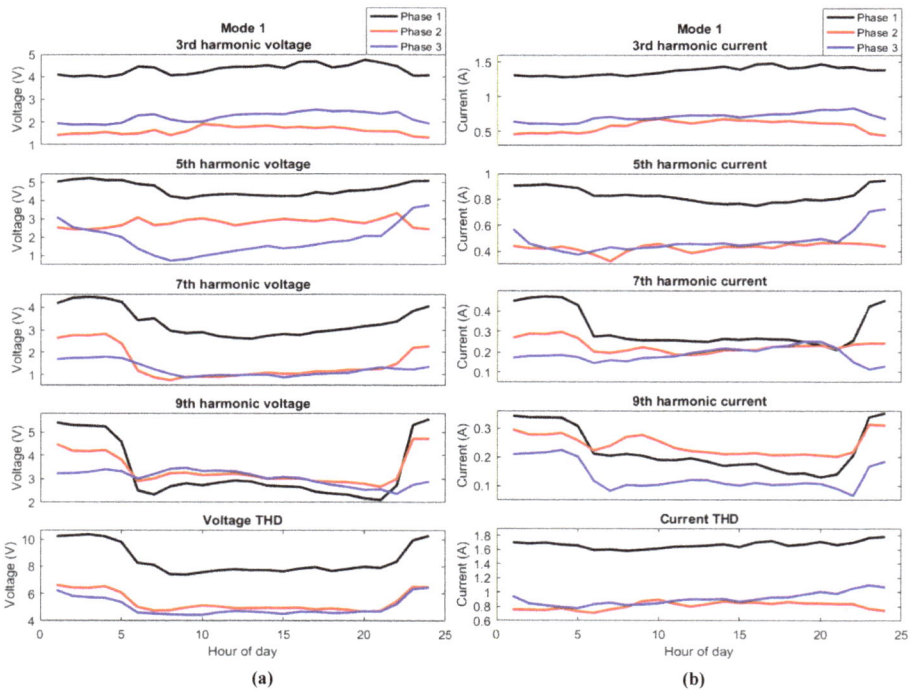

Figure 18. (**a**) The odd harmonic voltage 3rd to 9th and THD for mode 1 in Figures 14–16. (**b**) The odd harmonic current 3rd to 9th and THD for mode 1 in Figures 14–16. The average values are grouped for each hour of the day for around 8 weeks of islanded operation. Note also the difference in vertical scale for the different plots.

The average values for each hour during the day for around 8 weeks in islanded operation for the 3rd to 9th odd harmonic voltages, currents and THD for mode 2 in Figures 14–16 can be seen in Figure 19. The average 3rd harmonic voltage is larger during the middle of the day and with larger magnitude than in Figure 18. The 5th and 7th harmonic has the largest voltage distortion levels during the night for all 3 phases and larger variation between night and day in comparison to mode 1 seen in Figure 18. The 9th harmonic voltage is larger during the night for phase 3 and higher during the middle of the day for phase 1 and 2. The voltage THD has the largest values during the night which corresponds with the result from the longer 29 week measurements presented in Figure 3. The harmonic current levels follow the same pattern for some of the voltage harmonics but have opposite pattern for other voltage harmonics. This indicates that there are more factors that act on the voltage distortion than the current distortion.

In Figure 20, the 1 h average values are plotted for the 3rd and 5th harmonic current and voltage for around 8 weeks in islanded operation. The blue colored values are for mode 1 and the red colored values are for mode 2 seen in Figures 14–16. The inclination (slope) corresponds to the impedance which is higher for mode 2 than for mode 1. However, for mode 1 the values form a more linear behavior than for mode 2. For the 3rd harmonic in phase 3, there is an appearance of a large area that has variation in voltage distortion without an increase in current distortion magnitude.

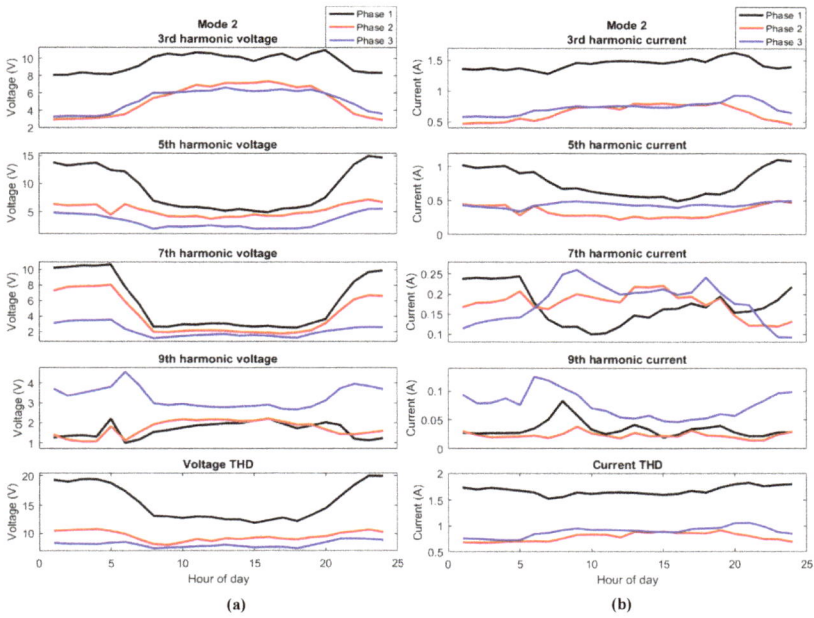

Figure 19. (**a**) The odd harmonic voltage 3rd to 9th and THD for mode 2 in Figures 14–16. (**b**) The odd harmonic current 3rd to 9th and THD for mode 2 in Figures 14–16. The average values are grouped for each hour of the day for around 8 weeks of islanded operation. Note also the difference in vertical scale for the different plots.

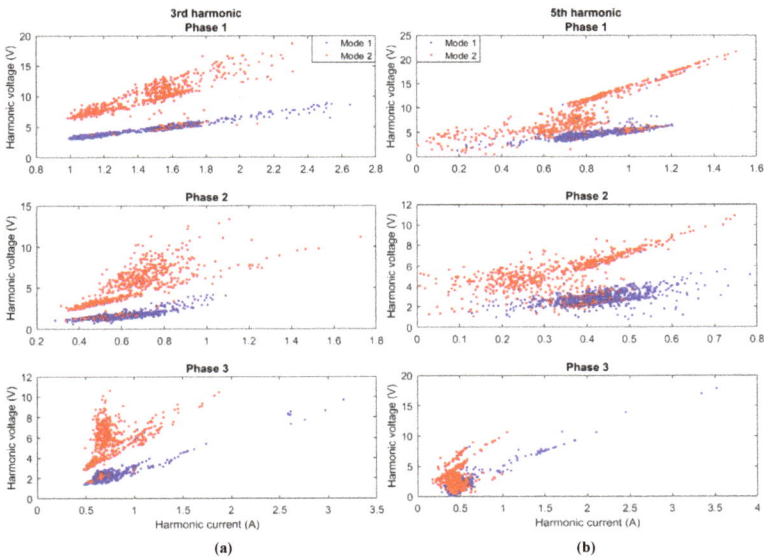

Figure 20. (**a**) The 1 h 3rd harmonic voltage and current values plotted against each other for Mode 1 and 2 seen in Figures 14–16. (**b**) The 1 h 5th harmonic voltage and current values plotted against each other for Mode 1 and 2 seen in Figures 14–16.

Another observation is that some of the measurements for mode 1 and 2 overlap each other which are expected since no criteria for the separation of the two modes exist except for the different time windows in Figures 13–16.

In Figure 21, the 1 h values are plotted for the 7th and 9th harmonic current and voltage for all three phases for around 8 weeks in islanded operation. The blue colored values are mode 1 and the red colored values are for mode 2 seen in Figures 14–16. The 7th and the 9th harmonic have some linear appearance for mode 1 but not as much as the 3rd and 5th harmonic. For mode 2 the behavior is more nonlinear, where some of the voltage distortion values are located close to zero current distortion for some phases. This indicates that the voltage distortion originates from the voltage source which is one of the possible sources of error in the impedance measurements. This is more evident in the 9th harmonic than in the 7th harmonic.

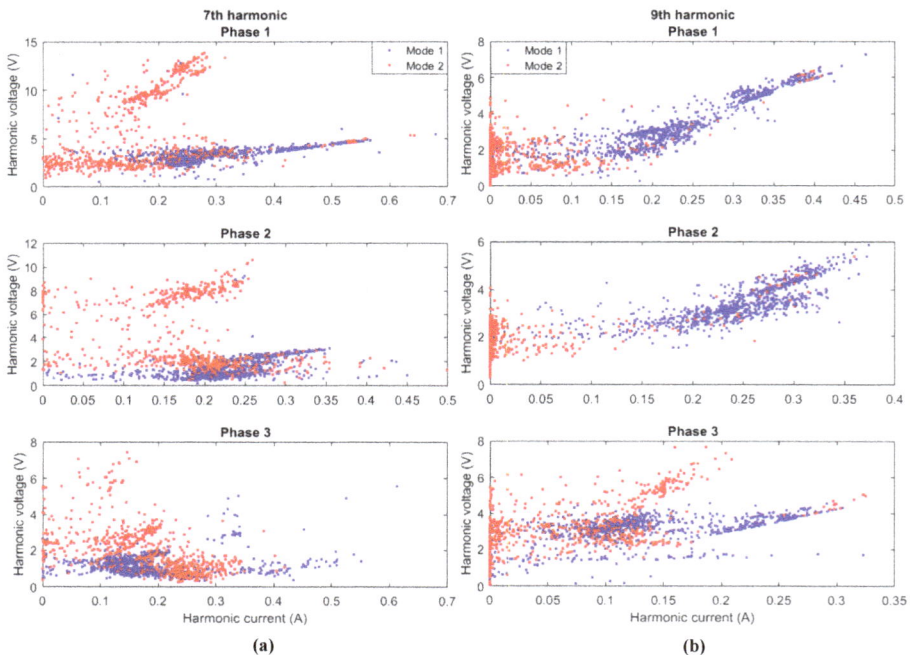

Figure 21. (**a**) The 1 h 7th harmonic voltage and current values plotted against each other for mode 1 and 2 seen in Figures 14–16. (**b**) The 1 h 9th harmonic voltage and current values plotted against each other for mode 1 and 2 seen in Figures 14–16.

4. Comparison to Reference Values

In this section, some of the measured values presented in Section 3 are compared to the European standard EN 50160 and the IEEE standard 519-2014.

4.1. Voltage THD

The allowed 10 min voltage THD for a low voltage networks is 8% for 95% of the time per week for both IEEE standard 519-2014 and European standard EN 50160. IEEE standard 519-2014 has also a 12% THD limit for 3 s values for 99% of the time for one day at the low voltage Point of Common Coupling (PCC).

During islanded operation the number of weeks that exceeded the 5% allowed time frame in which the 10 min voltage THD could be above 8% is shown in the upper part of Figure 22. The total number of days that the 3 s voltage THD was over 12% for more than 1% of one day can also be seen in the upper part of Figure 22. The amount of time in which the voltage THD was over 8% and 12% can be seen in the lower part in Figure 22. Only phase 1 surpassed the 5% weekly limit for 8% voltage THD in EN 50160 and the 1% daily limit for 12% voltage THD described in IEEE 519-2014. During grid-connected operation no limits was surpassed.

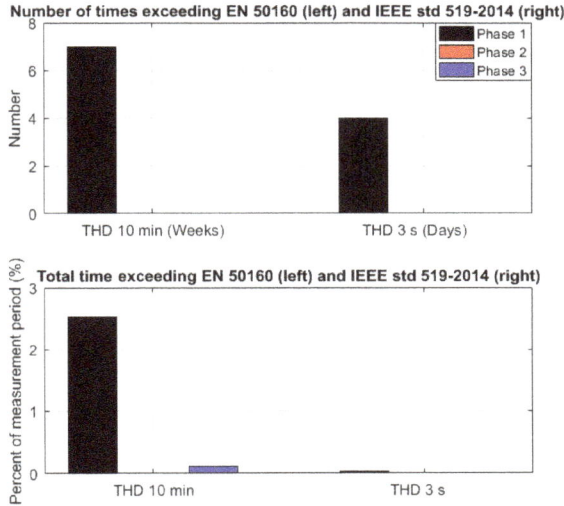

Figure 22. The number of times the voltage THD exceeded the limits in EN 50160 and IEEE 519-2014 for the 10 min (in weeks) and 3 s (in days) (**top**) and the amount of time in which the voltage THD exceeded the 8% and 12% limits in EN 50160 and IEEE 519-2014 (**bottom**) for all three phases during islanded operation. The measurement period was 48 weeks.

4.2. Individual Harmonics

For 95% of the time during one week the 10 min value of the individual harmonics shall not go beyond the limits in Table 9 from EN 50160. For IEEE 519-2014 the limit for individual harmonics is 5% for 10 min values and 7.5% for 3 s values at a low voltage PCC.

Table 9. EN 50160 limits for individual harmonics at the low voltage Point of Common Coupling (PCC).

Harmonic Order	Percent of Fund (%)
3rd	5
4th	1
5th	6
6th	0.5
7th	5
8th	0.5
9th	1.5
11th	3.5
15th	0.5

For both islanded and grid-connected operation, the number of weeks and total time in which the individual harmonics exceeded the limits in EN 50160 is shown in Figure 23. For islanded operation some individual harmonics stays within the 5% allowed time per week but exceed the limit sometime during the measurement period. Phase 2 during islanded operation never exceeds the 5% time limit. The 15th harmonic is only surpassed for one week for phase 3 during islanded operation. It can also be seen that the 9th harmonic exceeds almost every week for all three phases during islanded operation. For the grid-connected measurements, the 9th harmonic exceeded the 1.5% limit for 1 week.

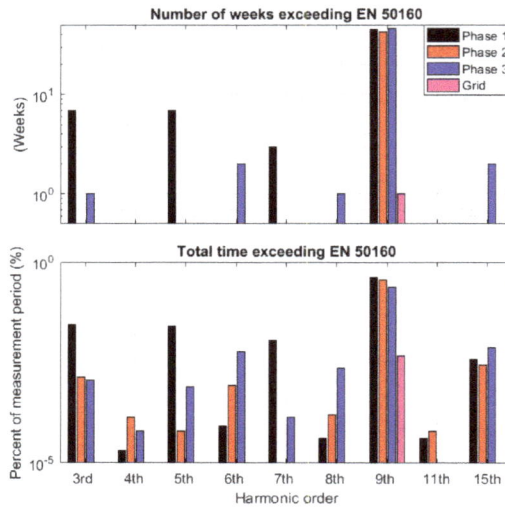

Figure 23. The number of times the individual harmonics exceeded the 10 min limits in EN 50160 (**top**) and the amount of time during the measurement period (48 weeks in islanded operation and 54 weeks in grid-connected operation) in which the individual harmonics exceeded the limits in EN 50160 (**bottom**) for islanded and grid-connected operation. Note the logarithmic vertical scale.

The number of weeks and amount of time in which the individual harmonics exceeded the limits in IEEE 519-2014 are shown in Figure 24.

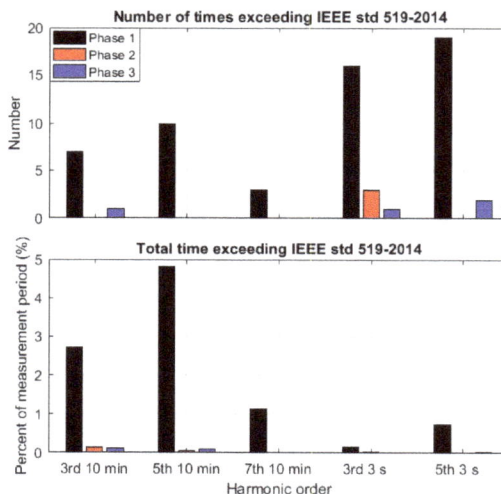

Figure 24. The number of times the individual harmonics exceeded the 10 min (in weeks) and 3 s limit (in days) in IEEE 519-2014 (**top**) and the amount of time during the measurement period of 48 weeks in which the individual harmonics exceeded the limits in IEEE 519-2014 (**bottom**) for islanded operation.

It can be seen that the 3rd, 5th and 7th harmonic are the only harmonics in the 10 min values that exceed the 5% limit in IEEE 519-2014 during islanded operation. For the 3 s values the 3rd and 5th harmonics are the ones exceeding the 7.5% limit during islanded operation. The limits were not surpassed during grid-connected operation.

4.3. Long Term Flicker Severity

In EN 50160 only the Plt value is specified which should be lower than 1 for 95% of the time for one week. For phase 1 and 3 the Plt value exceeded the limit for 3 weeks and phase 2 exceeded the limit for 4 weeks. The grid-connected measurements never exceeded the 5% allowed time limit in EN 50160. In Figure 25 the amount of time during the measurement period in which the Plt value exceeds the specified limit in EN 50160 for islanded and grid-connected is illustrated. The Plt values for islanded operation exceeded the limit in EN 50160 for about twice as long than in grid-connected operation.

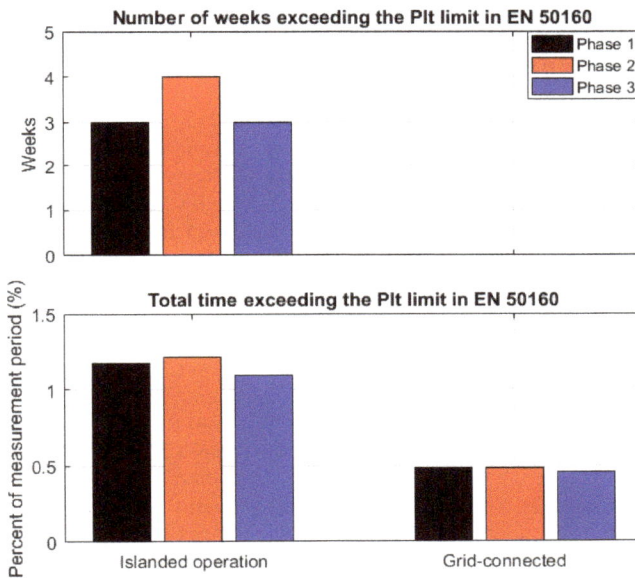

Figure 25. The number of times the Plt values exceeded the limits in EN 50160 (**top**) and the amount of time during the measurement period of 48 weeks in which the Plt values exceeded the limits in EN 50160 (**bottom**) for islanded operation.

4.4. Voltage Unbalance

According to EN 50160 the voltage unbalance should not go beyond 2% for more than 5% of the time for one week. In the grid-connected operation the voltage unbalance never went above 2%. In islanded operation the voltage unbalance went above 2% for 2.34 h, above 3% for 1.5 h in the 48 week measurements. The largest voltage unbalance during the longest consecutive time was 4.5% during 30 min. However, the voltage unbalance was kept below 2% for more than 95% of each week.

5. Discussion

Effects on Equipment

It is stated in [12] that for a voltage THD between 8 and 10% motors could get overheated. The measurements presented in Section 3 show that there can be levels that go beyond 8 and 10%

distortion in the nanogrid. This can lead to overheating for motors and other equipment sensitive to voltage harmonics.

Flat topping of the voltage waveform can affect switched-mode power supplies since they need a high peak voltage to effectively charge its capacitor. Since the 10 min voltage THD values can reach up to 13% with up to 9.3% of 3rd harmonic content, equipment that has switched-mode power supplies could also suffer from excess heating. Especially harmonics 5 and 7 can also cause more losses in electrical motors and a counter electromotive force that leads to torque and vibration problems. Fluctuations in the voltage can affect several types of equipment (excluding flicker). One of those effects is accelerating or braking torques in motors, deterioration of electronic equipment in which the voltage fluctuations pass through the power supply to the electronic equipment such as computers, printers, control units, components for telecommunication [13,14]. In a production facility, these fluctuations could lead to variations in the speed of motors that cause unacceptable variations in production parameters such as color and diameter. There is no standard today that defines acceptable levels of VSV. Since the majority VSV values are lower in islanded operation than for grid-connected operation, there is an improvement of the performance in islanded operation when comparing to grid-connected operation. However, more research is needed before any conclusion can be drawn on how equipment will perform under high VSV values.

The main effect of high Pst and Plt values is light flicker for incandescent lamps. Since the nanogrid only has LED lamps there could be either less or more light flicker depending on which LED lamp is used [15]. The difference in voltage flicker levels was larger between one phase and the other two phases for the majority of the time in islanded operation. This might suggest that the connected equipment might affect the flicker levels in a certain phase more in islanded operation than in grid operation since the flicker levels in the three phases during grid operation were closer to each other.

The winding temperature increase in percent above rated temperature as a function of the voltage unbalance can be calculated from Equation (1) [16]:

$$Temp\ rise\ = 2(Percent\ unbalance)^2 \tag{1}$$

From Equation (1) a temperature rise of 40.5% for 30 min could occur during the 48 week measurements for islanded operation. The maximum temperature rise would be 42.3% for 10 min. The average winding temperature increase for islanded operation is about 0.1% and about 0.2% for grid-connected operation. The reduction in lifetime of the motor due to temperature increase following the Arrhenius model is described in IEEE standard 101 [17]. Since the average winding temperature for a 3-phase motor is lower in islanded operation than in grid-connected operation, the lifetime of a 3-phase motor might increase in islanded operation. However, since there are times in which the motor can run on temperature elevations of up to 42.3% for shorter time periods at the same time as the voltage THD is above 8% (Mode 2 in Figure 14), a decrease in the lifetime of a motor could happen in comparison to grid-connected operation.

6. Conclusions

The voltage THD and individual harmonics reach larger values in islanded operation than in grid-connected operation. The difference in magnitude between the phases is also larger in islanded operation.

- Figure 5 shows that the voltage THD increases during the evening without a change in load which suggests that the system impedance increases due to fewer parallel sources being active. This could be due to the shutdown of the solar inverter in the evening. The subsequent decrease in voltage harmonics levels in the morning could be caused by the activation of the solar inverter.
- During islanded operation the Pst and Plt values were lower for the majority of the time for phase 2 while the other two are for the majority of the time higher in islanded operation.

- The 10 min and 3 s VSV value are for the majority of the time lower in islanded operation for all three phases. It can therefore be concluded that voltage variations on timescales 3 s to 10 min are lower in islanded operation.
- The voltage unbalance is lower for the majority of the time in islanded operation but could reach levels of over 4.5% for 30 consecutive minutes in islanded operation.
- Figures 13–21 show that there are two modes of performance during islanded operation. Mode 2 operates with larger values of the short circuit impedance, voltage THD, voltage unbalance and individual voltage harmonics 3, 5, 7 for all three phases. For phase 3, the 9th harmonic voltage was also larger during mode 2. Mode 1 operates with a larger 9th harmonic voltage for phase 1 and 2. The 3rd harmonic impedance behaves almost linear in both modes and is larger during mode 2 for all three phases. The 5th harmonic impedance behaves more nonlinear but is generally larger during mode 2 for two phases. The 7th and 9th harmonic impedance show nonlinear behavior and no conclusion can be made regarding the magnitude of the harmonic impedance for the 7th and 9th harmonic.
- When the islanded operation was compared to the limits defined in grid standard EN 50160 it was found that the voltage THD, individual harmonics 3, 5, 6, 7, 9, 15, voltage unbalance and Plt limits were exceeded. For one week the 9th harmonic exceeded the limit during grid-connected operation.
- When the islanded operation was compared to the limits defined in grid standard IEEE 519-2014 it was found that the voltage THD, individual harmonics 3, 5, 7 were exceeded. The limits were not exceeded during grid-connected operation.

More studies with long term measurements are needed to see if similar results are obtained for other nanogrids.

Author Contributions: Conceptualization, J.N., S.K.R. and M.H.J.B.; Methodology, J.N.; Software, J.N.; Validation, J.N., S.K.R. and M.H.J.B.; Formal Analysis, J.N.; Investigation, J.N.; Resources, S.K.R. and M.H.J.B.; Data Curation, J.N.; Writing-Original Draft Preparation, J.N.; Writing-Review & Editing, J.N., S.K.R. and M.H.J.B.; Visualization, J.N.; Supervision, S.K.R. and M.H.J.B.; Project Administration, S.K.R. and M.H.J.B.; Funding Acquisition, S.K.R. and M.H.J.B.

Funding: This paper has been funded by Skellefteå Kraft Elnät and Rönnbäret foundation.

Conflicts of Interest: The authors declare no conflict of interest.

References

1. Hatziargyriou, N. *Microgrids Architectures and Control*, 1st ed.; Wiley-IEEE Press: Hoboken, NJ, USA, 2014; pp. 310–313.
2. Hatziargyriou, N.; Asano, H.; Iravani, R.; Marnay, C. Microgrids. *IEEE Power Energy Mag.* **2007**, *5*, 78–94. [CrossRef]
3. CIGRÉ WG C6.22. *Microgrids 1: Engineering, Economics, & Experience*; The International Council on Large Electric Systems: Paris, France, 2015.
4. Burmester, D.; Rayudu, R.; Seah, W.; Akinyele, D. A review of nanogrid topologies and technologies. *Renew. Sustain. Energy Rev.* **2017**, *67*, 760–775. [CrossRef]
5. Cenelec Std. EN 50160:2010. *Voltage Characteristics of Electricity Supplied by Public Electricity Networks*; European Committee for Electrotechnical Standardization: Brussels, Belgium, 2010.
6. IEEE Std. 519-2014. *IEEE Recommended Practice and Requirements for Harmonic Control in Electric Power Systems*; IEEE Standards Association: Piscataway, NJ, USA, 2014.
7. Nömm, J.; Rönnberg, S.K.; Bollen, M.H.J. An Analysis of Frequency Variations and its Implications on Connected Equipment for a Nanogrid during Islanded Operation. *Energies* **2018**, *11*, 2456. [CrossRef]
8. Rönnberg, S.K.; Bollen, M.H.J.; Nömm, J. Power Quality Measurements In a Single House Microgrid. In Proceedings of the CIRED 24th International Conference on Electricity Distribution, Glasgow, Scotland, UK, 12–15 June 2017; pp. 818–822.

9. Nömm, J.; Rönnberg, S.K.; Bollen, M.H.J. Harmonic Voltage measurements in a Single House Microgrid. In Proceedings of the ICHQP 18th International Conference on Harmonics and Quality of Power, Ljubljana, Slovenia, 13–16 May 2018; pp. 1–5.

10. Bollen, M.H.J.; Häger, M.; Schwaegerl, C. Quantifying voltage variations on a time scale between 3 seconds and 10 minutes. In Proceedings of the CIRED 18th International Conference and Exhibition on Electricity Distribution, Turin, Italy, 6–9 June 2005; pp. 1–5.

11. Bollen, M.H.J.; Gu, I.Y.H. Characterization of voltage variations in the very-short time-scale. *IEEE Trans. Power Deliv.* **2005**, *20*, 1198–1199. [CrossRef]

12. Dugan, R.C.; McGranaghan, M.F.; Santoso, S.; Beaty, H.W. *Electrical Power System Quality*, 2nd ed.; McGraw-Hill Education: New York, NY, USA, 2003; p. 216.

13. Schlabbach, J.; Blume, D.; Stephanblome, T. *Voltage Quality in Electrical Power Systems*; The Institution of Electrical Engineers: Stevenage, UK, 2001; pp. 115–116.

14. UIE WG 2. *Guide to Quality of Electrical Supply for Industrial Installations, Part 5; Flicker and Voltage Fluctuations*; International Union for Electricity Applications: Paris, France, 1999; p. 13.

15. Gil-de-Castro, A.; Rönnberg, S.K.; Bollen, M.H.J. Light intensity variation (flicker) and harmonic emission related to LED lamps. *Electr. Power Syst. Res.* **2017**, *146*, 107–114. [CrossRef]

16. Pillay, P.; Manyage, M. Derating of Induction Motors Operating With a Combination of Unbalanced Voltages and Over or Undervoltages. *IEEE Trans. Energy Convers.* **2002**, *17*, 485–491. [CrossRef]

17. IEEE Std. 101-1987(R2010). *Guide for the Statistical Analysis of Thermal Life Test Data*; IEEE Standards Association: Piscataway, NJ, USA, 2010.

energies

MDPI

Article

Advantages of Minimizing Energy Exchange Instead of Energy Cost in Prosumer Microgrids

Eva González-Romera [1,*], Mercedes Ruiz-Cortés [1], María-Isabel Milanés-Montero [1], Fermín Barrero-González [1], Enrique Romero-Cadaval [1], Rui Amaral Lopes [2,3] and João Martins [2,3]

[1] Electrical, Electronic and Control Engineering Department, University of Extremadura, 06006 Badajoz, Spain; meruizc@peandes.es (M.R.-C.); milanes@unex.es (M.-I.M.-M.); fbarrero@unex.es (F.B.-G.); eromero@unex.es (E.R.-C.)
[2] Faculty of Sciences and Technology—NOVA, University of Lisbon, 2829516 Caparica, Portugal; rm.lopes@fct.unl.pt (R.A.L.); jf.martins@fct.unl.pt (J.M.)
[3] Center of Technology and Systems (CTS)—UNINOVA, 2829516 Caparica, Portugal
[*] Correspondence: evagzlez@unex.es; Tel.: +34-924-289-600

Received: 22 January 2019; Accepted: 19 February 2019; Published: 22 February 2019

Abstract: Microgrids of prosumers are a trendy approach within the smart grid concept, as a way to increase distributed renewable energy penetration within power systems in an efficient and sustainable way. Single prosumer individual management has been previously discussed in literature, usually based on economic profit optimization. In this paper, two novel approaches are proposed: firstly, a different objective function, relative to the mismatch between generated and demanded power, is tested and compared to classical objective function based on energy price, by means of a genetic algorithm method; secondly, this optimization procedure is applied to batteries' coordinated scheduling of all the prosumers composing a community, instead of single one, which better matches the microgrid concept. These approaches are tested on a microgrid with two household prosumers, in the context of Spanish regulation for self-consumption. Results show noticeably better performance of mismatch objective function and coordinated schedule, in terms of self-consumption and self-sufficiency rates, power and energy interchanges with the main grid, battery degradation and even economic benefits.

Keywords: distributed energy resources; electric energy storage; energy management system; genetic algorithm; microgrid; prosumer; self-consumption

---

1. Introduction

Within a context of reinforcing consumers' roles in the energy system and improving energy efficiency in Europe [1], the concept of prosumer has become common in the power system context. A prosumer is a consumer with the ability of producing electric energy to be either self-consumed or injected into the grid. A group of prosumers connected and coordinated to manage their energy resources are considered to be a microgrid [2]. The microgrid can be either isolated or grid-connected, in which energy management and economic profitability are the main targets.

Many studies can be found in literature analyzing microgrids' financial feasibility and the correct sizing of their respective Distributed Energy Resources (DERs) [3,4]. In most of the published works the main objective of the microgrid energy management system is to achieve the highest possible economic benefit by minimizing its operational costs [5,6]. Generation costs [6] and market prices, for the purchased/sold energy [5], are used to define optimization techniques' objective functions. However, economic incentives and additional costs have a great importance in the microgrids' financial sustainability and they must be considered in their design and operation [3]. Incentives are due to advantages of self-consumption improvement, that lie on empowering consumers by means of their participation in energy management programs, encouraging smarter consumption patterns, reducing peak demand and energy costs, decreasing transmission losses, due to the local generation and consumption, and contributing to finance energy transition [7].

Household prosumers usually count on locally available renewable energy generation, which, in most cases, is enabled by photovoltaic (PV) systems located at the dwellings [8]. In many countries, the decreasing cost of PV system devices has led to a grid-parity in self-generated PV energy in comparison to energy purchased from the grid. In this context, self-consumption achieves economic sustainability, even without financial support from governments.

Under deregulated electricity markets, extended all over Europe, general costs of power systems for building and operating infrastructures are covered by network tariffs and taxes. Network tariffs can be composed by two possible components: volumetric tariffs, according to which consumers pay per unit of purchased energy (€/kWh); and capacity tariffs, applied to the power capacity contracted with the energy provider (€/kW). In both cases tariffs can be classified as flat type, variable with consumption or capacity level and time-of-use type (different tariffs in peak/off-peak hours) [7]. Higher volumetric tariffs motivate for demand reduction, by means of energy efficiency and self-consumption. On the other hand, predominant capacity tariffs lead to peak-shaving.

Regarding energy prices, they are based on a spot market where a clearing price is obtained by the point at which the supply and demand curves meet, according to the European algorithm EUPHEMIA [9]. This market price establishes both purchase and sale prices for consumers and producers, respectively. In addition, the undeniable interest to increase the penetration of energy conversion systems based on renewable primary energy resources and the related motivation to improve self-consumption ratios has led to several promotion policies in different countries. An interesting classification of common incentives to introduce renewable energy resources can be found in [10]. According to this work, promotion policies can be focused on investment or on energy generation, but the most common type of policies in Europe are related to feed-in tariffs (FIT). There are two types of FIT systems: fixed and premium [7,10]. The first one is independent on the electricity market price, facilitating the assumption of investment risks, whereas the second one is an overprice above market price, with the advantage of being attached to the market behavior and thus avoiding excessive remuneration for producers and market distortion [7].

In the case of surplus energy fed into the grid, from prosumers using renewable energy resources, there is not a common understanding in the different European countries about the promotion policies. While many countries support investment or establish FIT frameworks, others prefer net metering approaches. Under this concept, the electricity excess injected into the grid can be later used during higher consumption hours, when self-generated energy is not sufficient [7]. Therefore, the main grid acts as a backup or a virtual storage system to manage the prosumers' energy surplus/deficit. Since energy price may be different during the hours of the day, and it is not considered in net-metering framework, this approach is commonly not favorable from the power system point of view and it is limited to small-size systems. As a consequence, some countries apply the so-called net-billing, which calculates the value of the energy fed into the grid at wholesale price to be used as a credit to purchase energy at other time period [7]. A survey of policies in European countries can be found in [7], although some of them have experienced variations during the last years.

Promotion policies make sense while immature technologies are trying to be introduced or while their marginal costs, when compared to conventional generation systems, do not allow them to participate in a competition market with equal conditions. The natural trend for any promotion policy is to converge into a market approach when grid-parity has been reached.

The case of Spain has been singular both in renewable energy, mainly PV, and in self-consumption promotion policies. The installation of renewable technologies and, especially, PV plants, was strongly boosted from 2007 to 2012, under high incentives in the form of fixed and premium FIT. In 2012 a high so-called "tariff-deficit" was reached which led to the extinction of support schemes for new plants and to changes in remuneration for existing plants, causing financial issues to many investors [11]. PV power plants were primarily promoted during the former years 2007 and 2008 which led to an extraordinary proliferation of this kind of generation systems and improved the experience and background of Spanish renewable energy companies all over the world. Since 2012, new PV power plants can only compete for extra economic support under a capacity-based aid to recover investment [12]. Regarding self-consumption, former regulation [13] stablished a specific cost to be paid as a consequence of self-consumed energy. Only low-voltage low-power prosumers who did not feed surplus energy into the grid were exempted from paying that specific cost. Additionally, self-consumption installations and energy could not be shared among prosumers, preventing the development of prosumers communities. This policy strongly discouraged the investment in new self-consumption installations. The situation has recently changed by means of a new law [14], which exempts self-consumed energy from paying any extra cost regardless of whether the prosumer feeds the excessive energy into the grid or not. Moreover, energy management can already be shared within a community. Therefore, in the absence of further regulation, current economic situation for Spanish prosumers can be summarized as the possibility of purchasing or selling energy from/to the power system at market price (no energy-based aids are planned for generated energy). In addition, payment of network tariffs for energy fed into or demanded from the grid must be paid, but no extra costs for self-consumed energy. Also, either individual or shared management are possible.

In order to maximize the advantage of self-consumption for end users, flexibility to match generated energy with demanded one is essential. This flexibility can be achieved by means of two possible mechanisms: demand-side response (DSR) and/or energy storage systems (ESS). The first one had a wide attention in literature [15,16]. As household electricity demand has a limited potential to apply restrictive DSR programs, the work described in this paper is focused on ESS usage instead, whose operation is scheduled according to different targets.

The role of ESS in microgrids is usually related to peak shaving [3], leading to a reduction of distribution system costs, which are based on the available power capacity instead of the volume of circulating energy. Other classical target is to mitigate fluctuation of PV generation [8] or other intermittent renewable energy source. ESS scheduling with economic target within a microgrid has also been discussed in literature [5,6,17,18]. Choi et al. [5] schedule a unique ESS to minimize traded energy cost based on energy price. In [6], the authors schedule batteries aiming at microgrid energy generation cost minimization, with no grid interchange. Both energy sale/purchase prices and operation and maintenance (O&M) costs of DERs are used in the optimization process in [17], although O&M costs of every DER (both renewable generation and ESS) are considered equal in this paper. Depreciation and environmental costs are also added to objective function in [18], being their values obtained from previous published papers. In all these works optimal ESS scheduling is performed and evaluated considering energy price as the only economic parameter to be considered in the interconnection of the microgrid with the main grid, i.e., to authors' knowledge, no previous work considers network tariffs to evaluate the economic advantages of an ESS scheduling method.

On the other hand, the coordination of several storage systems, owned by different prosumers, is not usually considered within a microgrid's common energy management strategy. The concept of "peer-to-peer (P2P)" supply [19,20] explains energy sharing within a community of prosumers. However, in these works, P2P supply is provided only by the generation surplus, without the involvement of stored energy. A wide review of possibilities of grouping prosumers as a community can be found in [21], which reveals that the common objectives of these communities are to maximize distributed production, to minimize losses or to optimize revenues. According to [21,22], further research is required in the area of prosumers' objectives and motivations.

The main goal of the present work is to evaluate the advantages of scheduling microgrid's ESS with an objective function based on minimizing generation-demand mismatch within the microgrid (thus minimizing energy exchange with the main grid) when compared to classical cost minimization. Benefits are observed for both prosumers, with PV generation and ESS, and the main grid. The case study is focused on a community of prosumers under Spanish regulation. Besides the expected advantages related to self-consumption, self-sufficiency and peak shaving, also economic benefits are achieved using the proposed objective function, because it reduces the prosumers electricity bill, at least in the absence of energy-based incentives, and better preserves battery lifetime. The economic evaluation of a scheduled optimization process that does not consider economic variables in the objective function is a novel contribution of this work. Furthermore, the study has been developed aiming to prove that cooperation between community prosumers, under the concept of microgrid, improves the global economic benefit for the community, compared to an individual management.

The paper is organized as follows: Section 2 describes the case study, the optimization methodology and the objective functions tested. Results are shown in Section 3. Finally, Section 4 discusses results and summarizes the conclusions.

2. Materials and Methods

2.1. Microgrid under Study

The microgrid considered in this study consists of two dwellings with typical household electrical devices, both equipped with a PV system and a battery based ESS (see Figure 1).

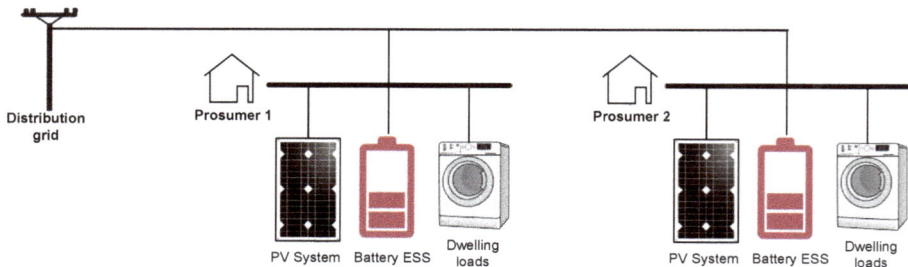

Figure 1. Schematic of the microgrid under study.

Starting from measured values in real households, demand of each house has been established considering typical residential appliances (lighting, air conditioning, washing machine, etc.) and a 4.6 kW peak power house (very common in Spanish dwellings). According to Spanish technical regulations the line capacity must be higher than the peak demand. House number 1 presents a higher demand during the 24 h of the day under study. Both houses are equipped with the same generation system and ESS (a 4 kW-rated power PV system and a battery with 6 kWh capacity and 2/−2 kW maximum charging/discharging power). Due to the proximity of the dwellings and the same characteristics of the PV systems, the same irradiation (based on climate data from a nearby weather station) and the same PV production has been assumed for both houses. Hourly average active power, generated and demanded, is shown in Figure 2 for both houses. The profiles from Figure 2 are used by the EES energy management system, assumed as a previous day forecast. One-hour resolution has been chosen not only for being an adequate resolution for one-day-horizon demand and generation forecasting, but also due to the time resolution of published market energy prices. In the scenario shown in Figure 2 one of the households has excess generated energy whereas the other one presents excess demand compared to the produced energy. Note that neither demanded power nor generated power exceeds peak demand (4.6 kW) and therefore, power lines' capacity is not exceeded.

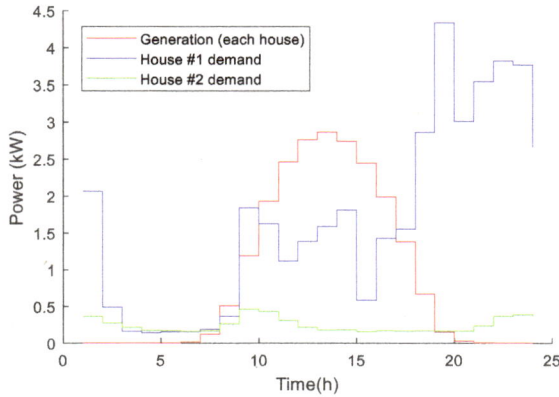

Figure 2. Generated and demanded average hourly power in both households.

2.2. Optimization Process for the Battery Scheduling

In this paper, the charging/discharging power of both batteries is scheduled with different target objectives, both in individual and in cooperative mode. The optimization process is performed using a genetic algorithm (GA) technique, as it was discussed in authors' previous work [23].

Generation and demand hourly power forecasted values (see Figure 2) of a prosumer are collected in 24-element vectors $G(t)$ and $D(t)$, $t = 1, 2, \ldots, 24$ h, respectively. In cooperative mode these vectors collect the sum of both prosumers. A 24-element vector $P_B^i(t)$, $t = 1, 2, \ldots, 24$ h, is composed of the dispatched active power schedule of prosumer's battery i. In cooperative mode all batteries within the community are considered. The adopted sign convention is positive for charging and negative for discharging. Elements $P_B^i(t)$ (hourly dispatched active power scheduled for every battery) are the values to be scheduled according to the optimization process; therefore, they will be the result of the GA technique. Original demand $D(t)$ is then modified to $D_{mod}(t)$, according to the battery contribution, as shown in Equation (1).

$$D_{mod}(t) = D(t) + \sum_i P_B^i(t). \tag{1}$$

Since batteries' charging/discharging efficiency rates present similar values, their effect has been neglected in this work.

The GA optimization process (described in [23]) finds the best solution for the elements $P_B^i(t)$. If an individual prosumer constitutes the objective to be optimized, the number of values to be obtained is 24. In a coordinated mode batteries of both prosumers are scheduled jointly, generating a vector composed of concatenated strings $P_B^1(t)$ and $P_B^2(t)$. This algorithm was programmed in MATLAB® to minimize the objective function subjected to a set of constraints, as explained below.

As previously pointed out, economic profit of prosumers is the most common optimization target addressed in literature. Energy cost is the preferred parameter to weight the scheduled battery power and, consequently, the purchased/sold energy from/to the main grid. Although this cost is not, by its own, sufficient to perform a complete optimization of the energy bill, the complexity of tariffs, premiums and prices encourages to adopt this value as weight for the objective function to be reached. As a consequence, in this work the price of traded energy is used to perform a first approach for the objective function. As discussed earlier, the purchased energy is usually paid at market prices, while the sold energy suffers from strong variations depending on the country. In Spain, currently, neither premiums nor fixed FIT are planned for energy sold by prosumers, therefore market price is also considered as the price for the sold energy. Final hourly price for each hour of an average day (average value for a day in a whole year, between 1 December 2017 and 30 November 2018 [24]) is collected in the 24-element vector $Pr(t)$ and shown in Figure 3.

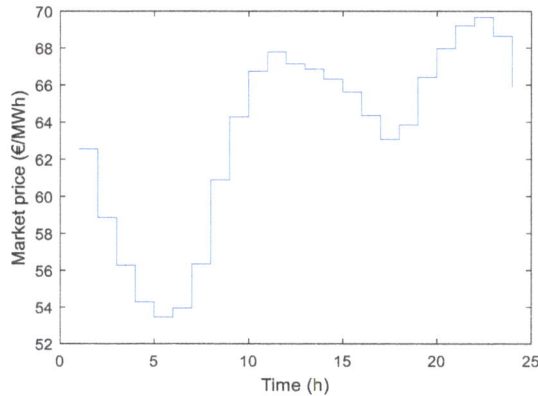

Figure 3. Spanish market yearly-average clearing price.

This preliminary approach for the objective function is called cost objective function (COF) and is determined by Equation (2), aiming to minimize energy cost where $f_{cost}(t)$ denotes the amount of money to be paid for the purchased energy minus the income received for the sold energy, at every hour $t = 1, 2, \ldots, 24$. The contribution of the batteries is included in the vector $D_{mod}(t)$, defined in Equation (1). Since PV generation and battery ESS O&M costs are not clearly related to their power scheduling, they have not been considered in the objective function:

$$f_{cost}(t) = Pr(t) \cdot [D_{mod}(t) - G(t)]. \tag{2}$$

In contrast, the novel objective function, proposed in this paper, is called mismatch objective function (MOF) and tries to improve not only the economic profit, but also the system efficiency (lower losses) and the exploitation of the DER. It intends to minimize the mismatch between generation and modified demand at every hour, i.e., the power in the interconnection with the main grid as shown in Equation (3). The square operator is introduced to penalize the highest errors. This objective function implicitly pursues self-consumption and self-sufficiency rates maximization, peak shaving, energy losses minimization in the distribution grid and network tariff costs reduction:

$$f_{mismatch}(t) = [D_{mod}(t) - G(t)]^2. \tag{3}$$

A set of constraints is added to the algorithm in order to assure safe performance and to prevent battery's early degradation. These constraints are the same in both optimization processes:

- State of charge (SoC) range: Battery manufacturers recommend keeping the SoC (percentage of charge related to the rated capacity) within a safe operational range $[SoC_{min}, SoC_{max}]$. Dynamic equation for SoC calculation is shown in Equation (4), under the assumption of unity charging/discharging efficiency. A discretization of Equation (4) is done in this constraint, which is shown in Equation (5), and minimum and maximum limits are assumed to be 20% and 100% respectively, taking [25] as reference:

$$SoC^i(t) = SoC^i(0) + \frac{1}{C_{nom}^i} \int_0^t P_B^i(t) dt, \tag{4}$$

$$SoC^i_{min} \leq SoC^i_{init} + \frac{\sum\limits_{t=1}^{24} P^i_B(t) \Delta t}{C^i_{nom}} \leq SoC^i_{max}. \tag{5}$$

For each battery i, C^i_{nom} is the nominal capacity (6000 Wh for both prosumers) and SoC^i_{init} is the SoC at the beginning of the day. In this study initial SoC of 83% and 50% have been considered for prosumers' batteries 1 and 2, respectively. The time interval Δt is 1 h.

- Charge/discharge power: The charging/discharging power is bounded by the battery specifications and by the battery's power electronics converter rated power. Equation (6) describes this constraint, in which $P^i_{max,dis}$ and $P^i_{max,ch}$ are the maximum allowed power for discharging and charging (-2000 W and 2000 W have been respectively assumed).

$$P^i_{max,dis} \leq P^i_B(t) \leq P^i_{max,ch}. \tag{6}$$

- Power gradient limitation: It is desirable that batteries' SoC presents a smooth variation along the day. Abrupt changes in SoC along with frequent alternating between charge and discharge modes negatively affect the battery lifetime [26,27]. Therefore, an additional constraint has been included to avoid large power oscillations between two consecutive hours. A maximum power gradient ΔP_B is considered with that purpose, as presented in Equation (7). A value of 300 W has been selected for ΔP_B, obtained from the maximum difference observed between consecutive hours in a daily profile averaged for a year in a real house.

$$\left| P^i_B(t) - P^i_B(t+1) \right| \leq \Delta P_B. \tag{7}$$

The ability of GA to search for the solution in the entire solution space and to find the global optimum solution even in non-convex, non-linear and non-smooth optimization problems has led to its use in this type of optimization problems [23,28]. This evolutionary algorithm emulates natural selection process. Starting from a random initial population of possible solutions (chromosomes) within a proper range (-2000 to 2000 W for each battery in this case study) of the solution space, GA iteratively creates new populations (children) according to their fitness for the objective function. Stopping criteria related to fitness value and computation time are included to finish the optimization procedure.

In this case study, chromosomes consist of 24 or 48 values of P_B, depending on whether individual or coordinated scheduling is being performed. The final solution will consist in the set of hourly battery charging/discharging power values which minimizes the objective function subject to defined constraints.

3. Results

The performance of the proposed algorithm with the aim of comparing the different objective functions and evaluating the advantages of prosumers' cooperation is quantified by means of usual indicators in literature. Self-consumption (SC) and self-sufficiency (SS) rates are well known indicators and they are presented in Equations (8) and (9), respectively. In both cases, the numerator computes the load instantly matched by the local production, using the considered batteries. Therefore, SC describes the percentage of on-site generation that is instantly consumed by the prosumer and SS the percentage of demand instantly satisfied by the on-site generation [29]:

$$SC = \frac{\sum\limits_{t=1}^{24} min[D_{mod}(t), G(t)]}{\sum\limits_{t=1}^{24} G(t)}, \tag{8}$$

$$SS = \frac{\sum\limits_{t=1}^{24} min[D_{mod}(t), G(t)]}{\sum\limits_{t=1}^{24} D(t)}. \tag{9}$$

Other technical indicators to evaluate the results are the power interchanged with the main grid P_{grid}, obtained from the nodal (in individual operation) or the whole system (coordinated operation) power balance, computed by Equation (10), and the accumulated energy import E_{imp} and export E_{exp} during the whole day, computed by Equations (11) and (12) respectively. One must note that P_{grid} should not be higher than the line capacity (4.6 kW for each dwelling or 9.2 kW for the common supply line):

$$P_{grid}(t) = D_{mod}(t) - G(t), \tag{10}$$

$$E_{imp} = \sum_{t=1}^{24} P_{grid}(t) \,\forall\, P_{grid}(t) > 0, \tag{11}$$

$$E_{exp} = \sum_{t=1}^{24} P_{grid}(t) \,\forall\, P_{grid}(t) < 0. \tag{12}$$

Finally, economic indicators are used to evaluate the profitability of the obtained solution for the prosumers. Based on the energy trading prices, $Cost_{bnt}$ denotes the energy cost before network tariffs in €/day, and is calculated by Equation (13):

$$Cost_{bnt} = \sum_{t=1}^{24} Pr(t) \cdot P_{grid}(t). \tag{13}$$

$Pr(t)$ is the set of hourly prices of the Spanish Electricity Market (Figure 3). According to P_{grid} sign, the cost is positive when the energy is imported.

Network tariffs have two components for consumers in Spain: a capacity-based term, Consumer Power Tariff (CPT, in €/kW·year), which is multiplied by the contracted power (CP), and a volumetric term, Consumer Energy Tariff (CET, in €/kWh). On the other hand, a volumetric term is only established for producers, Producer Energy Tariff (PET, in €/MWh). Values for these tariff terms have been obtained from Spanish regulation for low voltage consumers, with the so-called 2.0 A tariff [30], and for any producers [31], being CPT = 38.043426 €/kW·year, CET = 0.044027 €/kWh and PET = 0.5 €/MWh.

CP can be selected as any value of power, rounded by multiples of 0.1 kW, over the maximum demanded power [14]. In each case, CP (kW) has been obtained for the day under study, starting from the maximum absolute value of P_{grid}.

These values have been used to calculate the daily cost of the electric energy after network tariffs $Cost_{ant}$ by means of Equation (14):

$$Cost_{ant} = \text{CPT} \cdot \text{CP}/365 + \text{CET} \cdot |E_{imp}| + \text{PET} \cdot |E_{exp}|. \tag{14}$$

The mentioned indicators are used to evaluate the performance of both tested objective functions (cost objective function COF and mismatch objective function MOF), to compare them with each other and with the case without ESS. Firstly, individual scheduling for each dwelling is considered, and afterwards, results are compared to the case where coordinated community management is considered.

3.1. Individual Schedule of Batteries

Taking into account the objective functions presented in Section 2, Figure 4 shows generation, initial demand and demand modified by the action of the batteries for both prosumers.

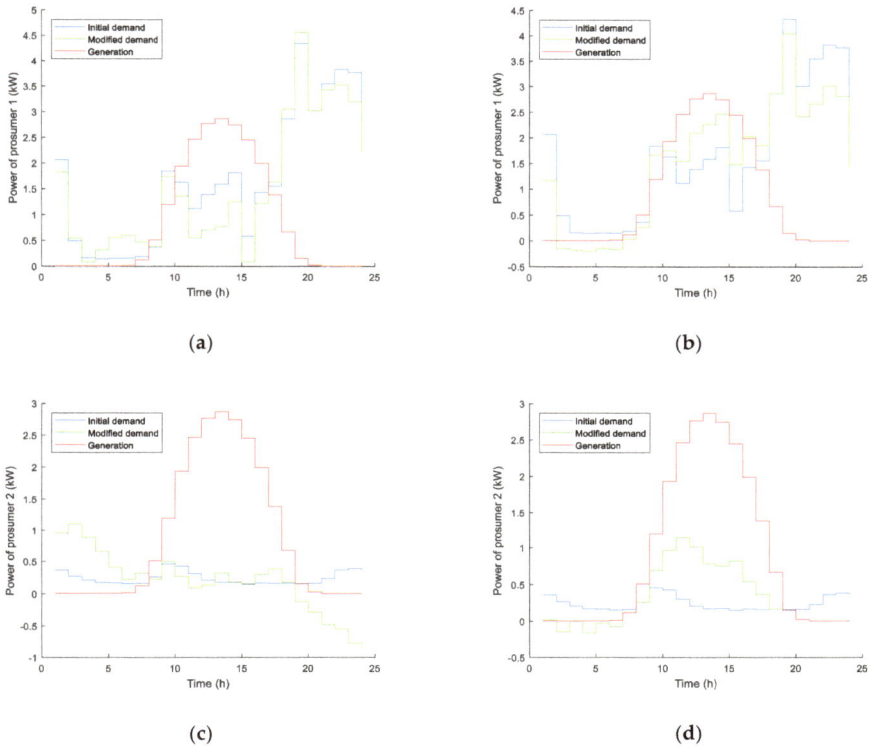

(a)

(b)

(c)

(d)

Figure 4. Generation, demand and modified demand hourly power curves during scheduled 24 h: (**a**) Prosumer 1, COF; (**b**) Prosumer 1, MOF; (**c**) Prosumer 2, COF; (**d**) Prosumer 2, MOF.

Table 1 shows the SC and SS indicators obtained with each of the proposed objective functions and the situation without ESS.

Table 1. SC and SS indicators for each of the prosumers with individual schedule.

Prosumer	Indicator	Without ESS	COF	MOF
Prosumer 1	SC	0.6327	0.4611	0.7568
	SS	0.3306	0.2410	0.3955
Prosumer 2	SC	0.1384	−0.0177 [1]	0.3427
	SS	0.5083	−0.0651 [1]	1.2582 [2]

[1] SS < 0 is due to negative values of $D_{mod}(t)$ in Equation (9). [2] SS > 1 means that imported energy is higher than household daily demand.

As it was expected, COF leads prosumers to buy energy at night (when the price is low) and to avoid buying or even to sell energy at evening peak price hours. This procedure has economic advantages in energy cost but penalizes SC and SS rates, as shown in Table 1. These indicators are even worse with COF than those obtained without ESS in the households. On the contrary, MOF significantly improves SC and SS of both prosumers, since its main target is to match generated and demanded power as much as possible.

Figure 5 shows the power interchanged with the main grid P_{grid}, according to Equation (10). It can be observed that line capacity of each dwelling is not exceeded in any cases.

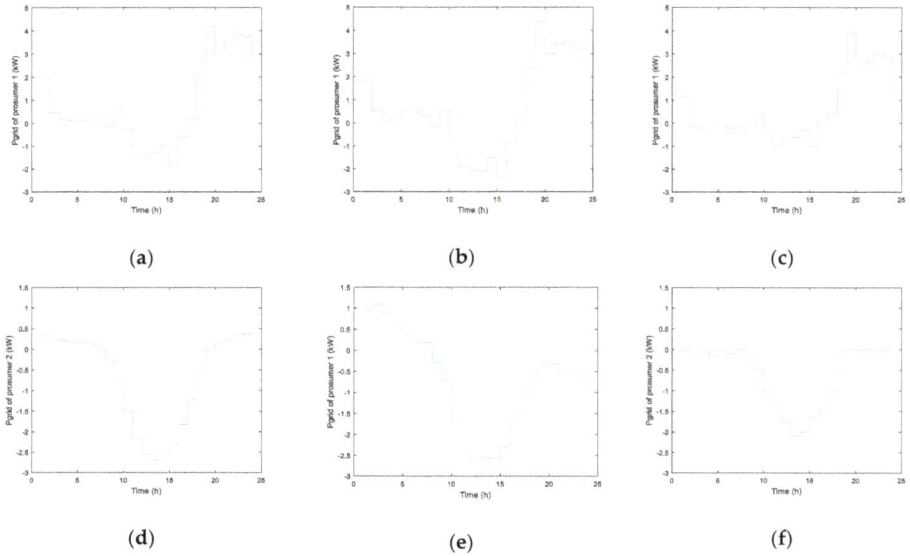

Figure 5. P_{grid} during scheduled 24 h (positive values for import energy flow and negative for export):
(**a**) Prosumer 1, without ESS; (**b**) Prosumer 1, COF; (**c**) Prosumer 1, MOF; (**d**) Prosumer 2, without ESS;
(**e**) Prosumer 2, COF; (**f**) Prosumer 2, MOF.

Absolute values for energy imported E_{imp} and exported E_{exp} (according to Equations (11) and (12)
respectively), are reported in Table 2.

Table 2. Energy import and export for each of the prosumers with individual schedule.

Prosumer	Indicator	Without ESS	COF	MOF
Prosumer 1	$\lvert E_{imp} \rvert$ (kWh)	27.203	27.047	20.495
	$\lvert E_{exp} \rvert$ (kWh)	7.801	11.445	4.893
Prosumer 2	$\lvert E_{imp} \rvert$ (kWh)	2.844	4.361	0.014
	$\lvert E_{exp} \rvert$ (kWh)	18.297	21.614	13.952

Figure 5 and Table 2 confirm the conclusion previously reached. The dependence of both
prosumers on the main grid can be, when COF is used, even higher than in the absence of ESS (either
in terms of power or energy). This is a consequence of considering energy price as the unique variable
to perform the battery management and leads to other drawbacks, such as higher required capacity of
the distribution grid and higher transmission losses. However, MOF reduces the dependence on the
main grid, both in power and energy terms, thus assuring advantages also for the distribution system.
The previous indicators used for evaluating the performance of the proposed objective functions
are related to technical performance. Table 3 below summarizes the economic indicators to evaluate
eventual prosumer benefits.

Table 3. Economic indicators for each of the prosumers with individual schedule.

Prosumer	Indicator	Without ESS	COF	MOF
	CP (kW)	4.2	4.4	3.9
Prosumer 1	$Cost_{bnt}$ (€/day)	1.2959	1.0258	1.0610
	$Cost_{ant}$ (€/day)	2.9351	2.6809	2.3723
	CP (kW)	2.7	2.6	2.1
Prosumer 2	$Cost_{bnt}$ (€/day)	−1.0278	−1.1826	−0.9135
	$Cost_{ant}$ (€/day)	−0.6120	−0.7088	−0.6871

Table 3 proves that MOF presents economic advantages as well. Although $Cost_{bnt}$ is obviously lower when COF is used, this approach leads to higher energy interchange with the grid, which adds costs in terms of network tariffs both when importing and exporting energy (although with different unitary cost). In the case of MOF, CP is reduced as well as imported and exported energy, leading to considerably lower network tariffs. As a consequence, this is the approach with the lowest $Cost_{ant}$ for prosumer 1 and produces similar revenues for prosumer 2.

Additional benefits are obtained from MOF related to batteries lifetime. As discussed earlier, a higher frequency of charge/discharge cycles increases battery degradation, leading to a lifetime decrease. Figure 6 compares batteries' SoC evolution when the different objective functions are used. It can be observed that oscillation between charge/discharge modes is less frequent using MOF than using COF, thus contributing to slow down battery degradation.

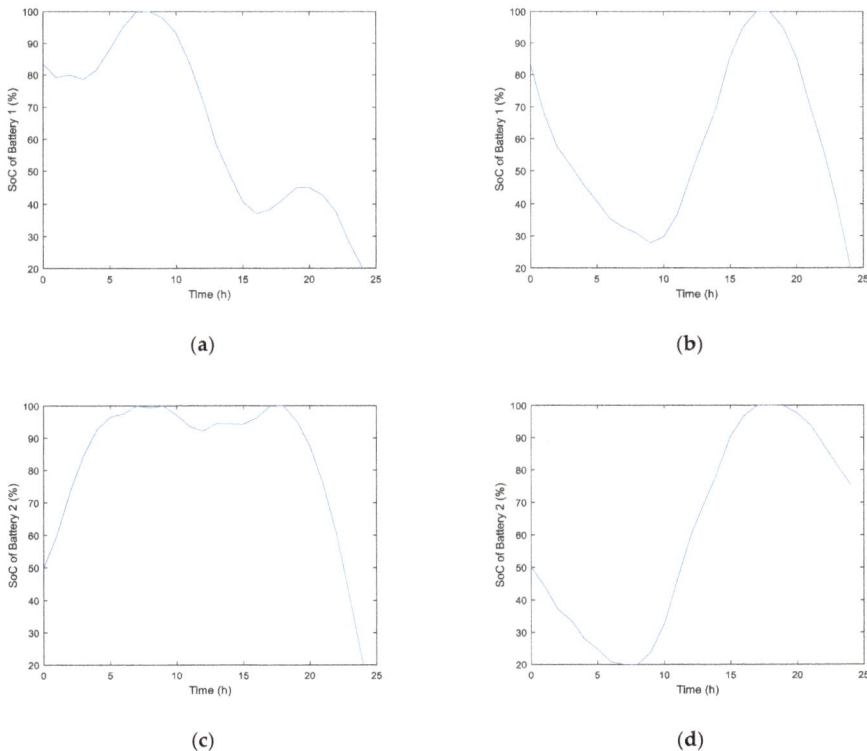

Figure 6. SoC of batteries during scheduled 24 h: (**a**) Prosumer 1, COF; (**b**) Prosumer 1, MOF; (**c**) Prosumer 2, COF; (**d**) Prosumer 2, MOF.

3.2. Coordinated Schedule of Batteries

At this stage, both prosumers are considered as one management unit (prosumers community), under the concept of microgrid, where DERs are shared to reach a joint optimization target. In this case, the output vector to be obtained, P_B, by means of the GA is composed of 48 values, corresponding to the 24 h scheduling of both batteries, while constraints are still applied to each of the batteries. The obtained results are shown below, with the aim of comparing this scenario to the individual management of each prosumer.

Figure 7 shows generation, initial demand and demand modified by the action of batteries, with both objective functions. It is important to note that these results refer to the whole microgrid.

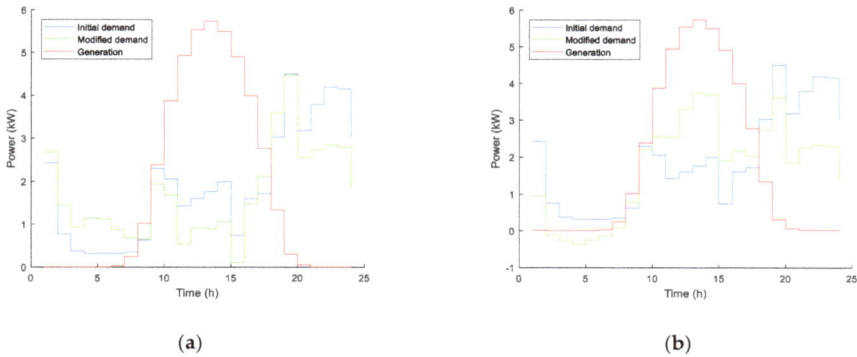

Figure 7. Generation, demand and modified demand hourly power curves of the whole microgrid, during scheduled 24 h: (**a**) COF; (**b**) MOF.

SC and SS indicators obtained for each objective function and for the situation without ESS can be found in Table 4. The same conclusions can be drawn as in the case of individual scheduling of batteries: MOF improves both SC and SS indicators, whereas they reach lower values when COF is used. This is due to a higher dependence on the main grid in this last case, as it is also shown in Figure 8 and Table 5.

Table 4. SC and SS indicators for the whole microgrid with coordinated scheduling.

Indicator	Without ESS	COF	MOF
SC	0.4168	0.3112	0.5950
SS	0.3814	0.2847	0.5444

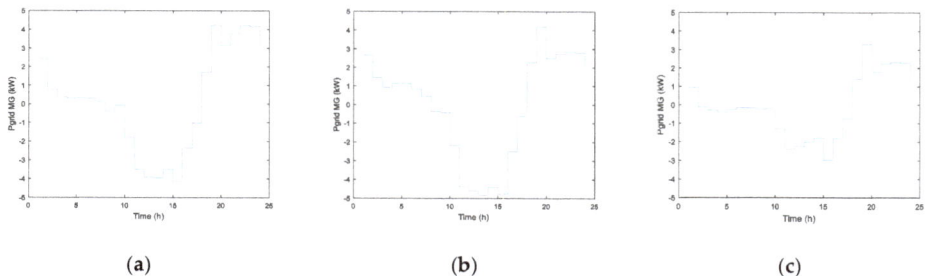

Figure 8. P_{grid} of the whole microgrid during scheduled 24 h (positive values for export energy flow and negative for import): (**a**) Without ESS; (**b**) COF; (**c**) MOF.

Table 5. Energy import and export for the whole coordinated microgrid.

Indicator	Without ESS	COF	MOF
$|E_{imp}|$ (kWh)	28.720	27.606	15.553
$|E_{exp}|$ (kWh)	24.771	29.257	17.204

Both power and energy interchange with the main grid are significantly improved when MOF is used. Note that P_{grid} is always far from the capacity of the common supply line (9.2 kW). In the case of COF, energy import is reduced in comparison to the case without ESS, but energy export is increased, since only economic factors are considered to reach the optimal solution.

The economic indicators are shown in Table 6, where the sum of both prosumers' costs with individual schedule has been added for comparison.

Table 6. Economic indicators for the whole microgrid, and individual CP and sum of both prosumers costs with individual schedule.

Type of Schedule	Indicator	Without ESS	COF	MOF
Coordinated (whole microgrid)	CP_{MG}(kW)	4.2	4.8	3.3
	$Cost_{bnt}$ (€/day)	0.2680	−0.1604	−0.0756
	$Cost_{ant}$ (€/day)	1.9826	1.5700	0.9617
Individual (sum of both prosumers)	CP_1/CP_2 (kW)	4.2/2.7	4.4/2.6	3.9/2.1
	$Cost_{bnt}$ (€/day)	0.2681	−0.1568	0.1475
	$Cost_{ant}$ (€/day)	2.3231	1.9721	1.6852

A very interesting conclusion can be drawn from Table 6. MOF also improves the economic profit of the microgrid in comparison to COF and moreover it significantly reduces the cost of the electricity bill for the whole microgrid compared to individual schedule of the battery of each prosumer. Obviously, the advantage is for the set of prosumers, as each of them is not equally affected. Prosumer 2 is a net energy exporter and earned money along the day (see Table 3). Therefore, if equitable share is done of the whole microgrid bill, this prosumer would be adversely affected. A fair rule based on energy balance of each prosumer should be developed to assign costs to each prosumer of the microgrid, as proposed in [21]. Regarding the evolution of battery SoC, Figure 9 depicts those regarding both objective functions, showing that MOF also reduces battery degradation.

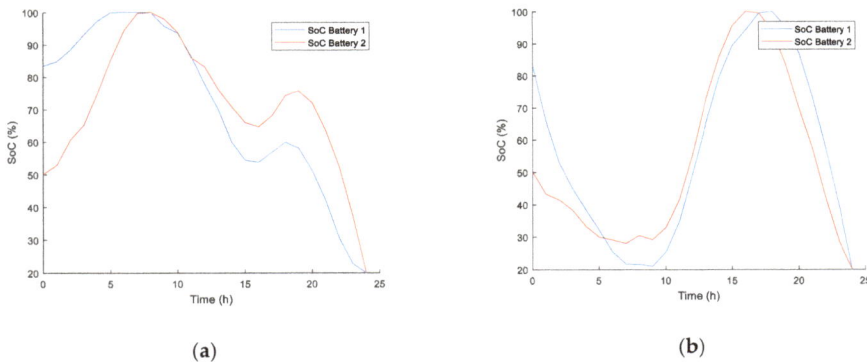

(a)   (b)

Figure 9. SoC of batteries during scheduled 24 h: (a) COF; (b) MOF.

The previously tested case study presented the advantages of batteries' coordinated scheduling in comparison to individual operation. In this case study, different demand levels have been considered to evaluate distinct performance between both operation modes. The proposed objective function was also tested using for both prosumers the average yearly demand profile obtained from a real measured dwelling. Figure 10 and Table 7 depict the generation, demand and modified demand of the whole microgrid and the main performance indicators obtained for this new case study, respectively.

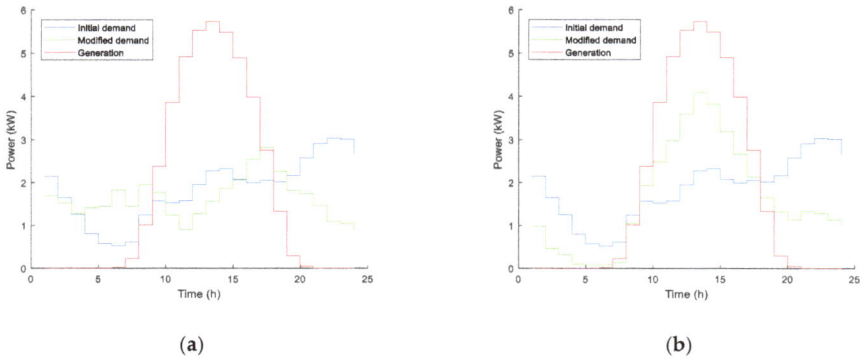

(a)　　　　　　　　　　　　　　　　　(b)

Figure 10. Generation, demand and modified demand hourly power curves of the whole microgrid, during scheduled 24 h, in the new case study: (a) COF; (b) MOF.

Table 7. SC and SS rates, energy import and export and economic indicators for the whole coordinated microgrid, in the new case study.

Indicators	Without ESS	COF	MOF
SC	0.4783	0.4456	0.7004
SS	0.4555	0.4244	0.6671
$|E_{imp}|$ (kWh)	24.285	20.071	9.248
$|E_{exp}|$ (kWh)	22.161	23.547	12.724
CP (kW)	3.6	4.2	1.9
$Cost_{bnt}$ (€/day)	0.1059	−0.3326	−0.2308
$Cost_{ant}$ (€/day)	1.5614	1.0006	0.3807

P_{grid} and the evolution of the SoC of batteries are also shown in Figures 11 and 12 respectively.

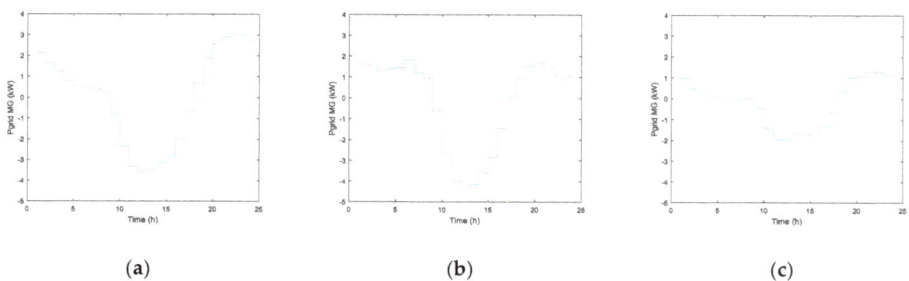

(a)　　　　　　　　　　(b)　　　　　　　　　　(c)

Figure 11. P_{grid} during scheduled 24 h (positive values for export energy flow and negative for import), in the new case study: (a) Without ESS; (b) COF; (c) MOF.

This new case study demonstrates that the proposed objective function, in a measured average demand situation, clearly outperforms the objective function based on energy cost in all the parameters previously discussed. It improves SC and SS indicators, significantly reduces the interchanged power and energy, decreases electricity bill and better exploits battery capacity, reducing the number of charging/discharging cycles a day.

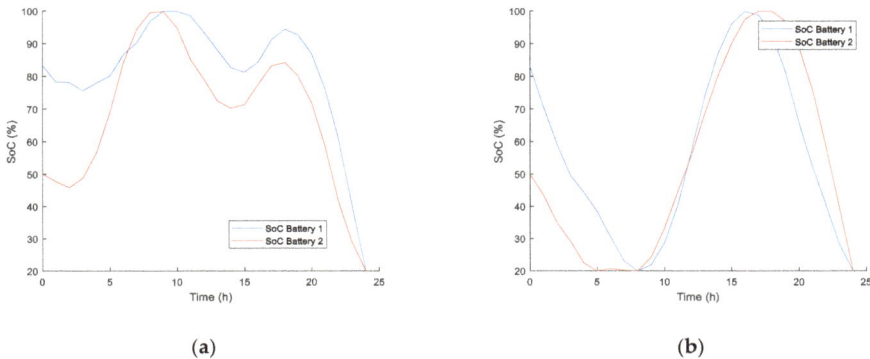

(a)   (b)

Figure 12. SoC of batteries during scheduled 24 h, in the new case study: (**a**) COF; (**b**) MOF.

4. Discussion and Conclusions

In this paper a microgrid of prosumers with PV generation and battery ESS is managed using a GA to schedule the batteries' charging/discharging power in a daily timeframe with one-hour resolution. Two management strategies were tested: individual and coordinated scheduling. Moreover, two objective functions were evaluated and compared. The first one (COF) aims to minimize the cost associated with energy consumption, while the second one (MOF) is focused on improving both SC and SS indices.

Regarding objective functions, the proposed MOF outperforms COF, which is based on traded energy price. This improvement is mainly reached in terms of SC and SS rates, and in the interconnection power and energy with the main grid. As a consequence, a better exploitation of DER is achieved, along with a lower impact into the distribution grid, both in capacity (which must be enough to match peak power) and in losses (which are related to the volume of flowing energy). But also economic advantages are shown when this new objective function is used comparing to an objective function based on the cost of the traded energy. Although COF leads to lower energy cost, before network tariffs, it also causes an increase in the required CP and in the daily imported and exported energy, thus increasing networks tariffs. This conclusion is valid in the absence of support for surplus energy of self-consumers, which tends to be the policy in European countries once grid-parity for the activity has been achieved. Finally, battery cycling is favorable to avoid premature degradation, as only one charge/discharge cycle is completed during the day with MOF, unlike the result of using COF. Results obtained from this study conclude that reducing the dependence on the main grid, both in power and energy terms, not only improves energy efficiency, distribution losses, exploitation of DERs and empowers consumers, but also economic advantages are reached for prosumers both in the electricity bill and battery lifetime. Therefore, this objective outperforms classical cost minimization techniques widely proposed in literature.

Another important conclusion is attained related to microgrid coordination. Performing the schedule of batteries as a set of resources to jointly optimize the microgrid energy management shows a noticeable improvement compared to individual scheduling, in terms of power and energy interchanges as well as economic benefits. Coordinated schedule of prosumers with average demand also shows clearly higher benefits when MOF is used instead of COF. This case study represents a realistic self-consumption situation and conclusions can be easily extended to bigger communities with several household prosumers. This result reinforces the interest for microgrids and smart communities as a way to improve energy efficiency and exploitation of DER. An effective method for assigning the resulting total costs of the microgrid to each of the prosumers is pending for future research.

Author Contributions: Conceptualization, E.G.-R. and F.B.-G.; Formal analysis, E.G.-R. and E.R.-C.; Funding acquisition, E.G.-R., M.-I.M.-M., F.B.-G. and E.R.-C.; Investigation, E.G.-R. and M.R.-C.; Methodology, M.R.-C. and J.M.; Project administration, E.G.-R.; Software, M.R.-C. and R.A.L.; Supervision, M.-I.M.-M., F.B.-G. and E.R.-C.; Validation, E.G.-R. and M.-I.M.-M.; Writing—original draft, E.G.-R.; Writing—review & editing, M.R.-C. and R.A.L.

Funding: This research was funded by the Spanish Agencia Estatal de Investigación (AEI), Fondo Europeo de Desarrollo Regional (FEDER), grant number TEC2016-77632-C3-1-R (AEI/FEDER, UE), Junta de Extremadura (Regional Government), Spain, under the Mobility Scholarship Program for Teaching and Research Staff of the Autonomous Community of Extremadura 2018 and Portugal national funds through FCT-Fundacão para a Ciência e a Tecnologia, under project UID/EEA/00066/2013.

Conflicts of Interest: The authors declare no conflict of interest.

Nomenclature

CET:	Consumer Energy Tariff
C^i_{nom}:	Nominal capacity of battery i
COF:	Cost objective function
$Cost_{ant}$:	Cost after network tariffs
$Cost_{bnt}$:	Cost before network tariffs
CP:	Contracted Power
CP_i:	Contracted Power of prosumer i
CP_{MG}:	Contracted Power of the whole microgrid
CPT:	Consumer Power Tariff
$D(t)$:	Demanded power at hour t
DER:	Distributed Energy Resource
$D_{mod}(t)$:	Modified demand at hour t
DSR:	Demand-Side Response
E_{exp}:	Exported energy along the whole day
E_{imp}:	Imported energy along the whole day
ESS:	Energy Storage System
f_{cost}:	Fitness function to be minimized according to COF
FIT:	Feed-in Tariffs
$f_{mismatch}$:	Fitness function to be minimized according to MOF
$G(t)$:	Generated power at hour t
GA:	Genetic Algorithm
MOF:	Mismatch objective function
O&M:	Operation and Maintenance
P2P:	Peer-to-Peer
$P_B^i(t)$:	Battery power of battery i at hour t
PET:	Producer Energy Tariff
P_{grid}:	Power in the interconnection with the main grid
$P^i_{max,ch}$:	Maximum charge power for battery i
$P^i_{max,dis}$:	Maximum discharge power for battery i
$Pr(t)$:	Energy market price at hour t
PV:	Photovoltaic
SC:	Self-Consumption rate
SoC:	State of Charge
SoC^i_{init}:	State of Charge of battery i at the beginning of the day
SoC^i_{max}:	Maximum allowed State of Charge of battery i
SoC^i_{min}:	Minimum allowed State of Charge of battery i
SS:	Self-Sufficiency rate
ΔP_B:	Maximum power gradient between consecutive hours

References

1. European Commission. The Strategic Energy Technology (SET) Plan. At the heart of Energy Research & Innovation in Europe. Available online: https://setis.ec.europa.eu/sites/default/files/setis%20reports/2017_set_plan_progress_report_0.pdf (accessed on 21 December 2018).
2. Albaker, A.; Khodaei, A. Elevating prosumers to provisional microgrids. In Proceedings of the 2017 IEEE Power & Energy Society General Meeting, Chicago, IL, USA, 16–20 July 2017; pp. 1–5. [CrossRef]
3. Hau, V.B.; Husein, M.; Chung, I.-Y.; Won, D.-J.; Torre, W.; Nguyen, T. Analyzing the Impact of Renewable Energy Incentives and Parameter Uncertainties on Financial Feasibility of a Campus Microgrid. *Energies* **2018**, *11*, 2446. [CrossRef]
4. Scalfati, A.; Iannuzzi, D.; Fantauzzi, M.; Roscia, M. Optimal sizing of distributed energy resources in smart microgrids: A mixed integer linear programming formulation. In Proceedings of the 2017 IEEE 6th International Conference on Renewable Energy Research and Applications (ICRERA), San Diego, CA, USA, 5–8 November 2017; pp. 568–573. [CrossRef]
5. Choi, J.; Shin, Y.; Choi, M.; Park, W.; Lee, I. Robust Control of a Microgrid Energy Storage System using Various Approaches. *IEEE Trans. Smart Grid*. In press. [CrossRef]
6. Luna, A.C.; Diaz, N.L.; Graells, M.; Vasquez, J.C.; Guerrero, J.M. Mixed-Integer-Linear- Programming-Based Energy Management System for Hybrid PV-Wind-Battery Microgrids: Modeling, Design, and Experimental Verification. *IEEE Trans. Power Electron.* **2017**, *32*, 2769–2783. [CrossRef]
7. European Commission. Best practices on Renewable Energy Self-consumption. Available online: https://ec.europa.eu/energy/sites/ener/files/documents/1_EN_autre_document_travail_service_part1_v6.pdf (accessed on 11 January 2019).
8. Kharseh, M.; Wallbaum, H. How Adding a Battery to a Grid-Connected Photovoltaic System Can Increase its Economic Performance: A Comparison of Different Scenarios. *Energies* **2019**, *12*, 30. [CrossRef]
9. PCR. EUPHEMIA: Description and Functioning. Available online: https://www.epexspot.com/document/40114/Euphemia%20Public%20Documentation (accessed on 17 January 2019).
10. Milanés-Montero, P.; Arroyo-Farrona, A.; Pérez-Calderón, E. Assessment of the Influence of Feed-In Tariffs on the Profitability of European Photovoltaic Companies. *Sustainability* **2018**, *10*, 3427. [CrossRef]
11. Roldan-Fernandez, J.M.; Gómez-Quiles, C.; Merre, A.; Burgos-Payán, M.; Riquelme-Santos, J.M. Cross-Border Energy Exchange and Renewable Premiums: The Case of the Iberian System. *Energies* **2018**, *11*, 3277. [CrossRef]
12. Ministerial Order ETU/315/2017 of 6 April, which Regulates the Procedure for Assignment Specific Remuneration Regime in the Call for New Electric Energy Production Installations Using Renewable Resources, Based of Royal Decree 359/2017, 31th March, and Sanctions Remuneration Parameters. Available online: https://www.boe.es/diario_boe/txt.php?id=BOE-A-2017-3880 (accessed on 14 January 2019).
13. Royal Decree 900/2015 of 9 October, which Regulates Administrative, Technical and Economic Conditions of Energy Supply with Self-Consumption and Energy Generation with Self-Consumption. Available online: http://www.omie.es/files/r.d._900-2015.pdf (accessed on 14 January 2019).
14. Royal Decree-Law 15/2018 of 5 October, of Urgent Actions for Energy Transition and Consumer Protection. Available online: http://www.omie.es/files/r.d.l._15-2018de_5_de_octubre_0.pdf (accessed on 14 January 2019).
15. Arif, A.; Wang, Z.; Wang, J.; Mather, B.; Bashualdo, H.; Zhao, D. Load Modeling—A Review. *IEEE Trans. Smart Grid* **2018**, *9*, 5986–5999. [CrossRef]
16. Tayade, A.; Dhote, V.P.; Thosar, A. Study on demand side management in microgrid using electric spring. In Proceedings of the 2018 International Conference on Current Trends towards Converging Technologies (ICCTCT), Coimbatore, India, 1–3 March 2018; pp. 1–6. [CrossRef]
17. Eseye, A.T.; Zheng, D.; Li, H.; Zhang, J. Grid-price dependent optimal energy storage management strategy for grid-connected industrial microgrids. In Proceedings of the 2017 Ninth Annual IEEE Green Technologies Conference (GreenTech), Denver, CO, USA, 29–31 March 2017; pp. 124–131. [CrossRef]
18. Liu, C.; Wang, X.; Wu, X.; Guo, J. Economic Scheduling Model of Microgrid Considering the Lifetime of Batteries. *IET Gener. Transm. Distrib.* **2017**, *11*, 759–767. [CrossRef]
19. Bellekom, S.; Arentsen, M.; van Gorkum, K. Prosumption and the Distribution and Supply of Electricity. *Energy. Sustain. Soc.* **2016**, *6*, 1–17. [CrossRef]

20. Zame, K.K.; Brehm, C.A.; Nitica, A.T.; Richard, C.L.; Schweitzer, G.D. Smart Grid and Energy Storage: Policy Recommendations. *Renew. Sustain. Energy Rev.* **2018**, *82*, 1646–1654. [CrossRef]

21. Espe, E.; Potdar, V.; Chang, E. Prosumer Communities and Relationships in Smart Grids: A Literature Review, Evolution and Future Directions. *Energies* **2018**, *11*, 2528. [CrossRef]

22. Milanés-Montero, M.-I.; Barrero-González, F.; Pando-Acedo, J.; González-Romera, E.; Romero-Cadaval, E.; Moreno-Muñoz, A. Active, Reactive and Harmonic Control for Distributed Energy Micro-Storage Systems in Smart Communities Homes. *Energies* **2017**, *10*, 448. [CrossRef]

23. Ruiz-Cortés, M.; González-Romera, E.; Amaral-Lopes, R.; Romero-Cadaval, E.; Martins, J.; Milanés-Montero, M.I.; Barrero-González, F. Optimal Charge/Discharge Scheduling of Batteries in Microgrids of Prosumers. *IEEE Trans. Energy Convers.* **2018**. [CrossRef]

24. OMIE. Iberian electricity Market Report. Available online: http://www.omie.es/reports/ (accessed on 2 January 2019).

25. Zhao, J.; Liu, N.; Lei, J. Co-benefit and profit sharing model for operation of neighboring industrial PV prosumers. In Proceedings of the 2015 IEEE Innovative Smart Grid Technologies—Asia (ISGT ASIA), Bangkok, Thailand, 17–20 February 2015; pp. 1–6. [CrossRef]

26. Zeh, A.; Müller, M.; Naumann, M.; Hesse, H.; Jossen, A.; Witzmann, R. Fundamentals of Using Battery Energy Storage Systems to Provide Primary Control Reserves in Germany. *Batteries* **2016**, *2*, 29. [CrossRef]

27. Uddin, K.; Gough, R.; Radcliffe, J.; Marco, J.; Jennings, P. Techno-Economic Analysis of the Viability of Residential Photovoltaic Systems Using Lithium-Ion Batteries for Energy Storage in the United Kingdom. *Appl. Energy* **2017**, *206*, 12–21. [CrossRef]

28. Mahesh, A.; Sandhu, K.S. A Genetic Algorithm Based Improved Optimal Sizing Strategy for Solar-Wind-Battery Hybrid System Using Energy Filter Algorithm. *Front. Energy* **2017**, 1–13. [CrossRef]

29. Quoilin, S.; Kavvadias, K.; Mercier, A.; Pappone, I.; Zucker, A. Quantifying Self-Consumption Linked to Solar Home Battery Systems: Statistical Analysis and Economic Assessment. *Appl. Energy* **2016**, *182*, 58–67. [CrossRef]

30. Ministerial Order ETU/1282/2017 of 22 December, which Establishes Electric Energy Network Tariffs for 2018. Available online: https://www.boe.es/buscar/pdf/2017/BOE-A-2017-15521-consolidado.pdf (accessed on 16 January 2019).

31. Royal Decree 1544/2011 of 31 October, Which Establishes Transmission and Distribution Network Tariffs That Must Be Satisfied by Electric Energy Producers. Available online: http://www.omie.es/files/r.d_1544-2011_de_31_de_octubre.pdf (accessed on 16 January 2019).

energies

MDPI

Article

An Alternative Internet-of-Things Solution Based on LoRa for PV Power Plants: Data Monitoring and Management

José Miguel Paredes-Parra [1], Antonio Javier García-Sánchez [2] and Antonio Mateo-Aroca [3] and Ángel Molina-García [3,*]

[1] Technological Center for Energy and Environment (CETENMA), 30353 Cartagena, Spain; jmparedes@cetenma.es

[2] Department of Information and Communication Technologies, Universidad Politécnica de Cartagena, 30202 Cartagena, Spain; antoniojavier.garcia@upct.es

[3] Department of Electronic Technology, Universidad Politécnica de Cartagena, 30202 Cartagena, Spain; antonio.mateo@upct.es

* Correspondence: angel.molina@upct.es; Tel.: +34-968-32-5462

Received: 12 January 2019; Accepted: 2 March 2019; Published: 6 March 2019

Abstract: This paper proposes a wireless low-cost solution based on long-range (LoRa) technology able to communicate with remote PV power plants, covering long distances with minimum power consumption and maintenance. This solution includes a low-cost open-source technology at the sensor layer and a low-power wireless area network (LPWAN) at the communication layer, combining the advantages of long-range coverage and low power demand. Moreover, it offers an extensive monitoring system to exchange data in an Internet-of-Things (IoT) environment. A detailed description of the proposed system at the PV module level of integration is also included in the paper, as well as detailed information regarding LPWAN application to the PV power plant monitoring problem. In order to assess the suitability of the proposed solution, results collected in real PV installations connected to the grid are also included and discussed.

Keywords: PV monitoring; low-cost solutions; LoRa technology

1. Introduction

In most countries, fossil fuel consumption has been drastically increasing along with enhancements in the quality of life and industrialization, and a growing world population [1]. This relevant fossil fuel consumption not only leads to an increase in the rate of diminishing fossil fuel reserves, but also has a significant adverse influence on the environment and the threat of global climate change. Actually, renewable integration issues have drawn attention in the scientific literature lately, and recent contributions have been focused on the institutional challenges [2]. Within the electricity sector, renewable and clean power generation alternatives will play a relevant role in future power supply (i) to attain global public awareness and sensibility of the need for environmental protection, and (ii) to achieve less dependence on fossil fuels for energy production [3]. Indeed, Mancarella et al. affirm that power systems are among the most critical infrastructures of modern societies, being crucially important to boost their resilience under severe weather conditions and any future challenges focused on climate change concerns [4]. As a result, most road maps and scenarios forecast a relevant resurgence of low-carbon generator units in the electricity supply side mix [5]. From the different renewable resources, wind and PV solar solutions are considered as relatively mature technologies, with a significant impact on current power systems [6]. However, certain technical problems have been discussed in the literature about high penetration of renewables, mainly focused on reliability, power quality, and stability [7]. In this way, the intermittent nature of such sources may increase the stress of the grid, mainly due to undesirable oscillations on the supply side [8], which may negatively affect the transmission system regulation [9]. Regarding PV power plants, their power generation is highly dependent on solar irradiance, ambient temperature, and other atmospheric parameters [10]. Consequently, fluctuations from grid-connected PV installations might lead to decreased grid reliability, compromising the demand–supply balance control [11]. Indeed, PV systems extensively integrated into low-voltage (LV) distribution grids may cause significant changes in feeder voltage profiles [12]. Consequently, it is very important to determine and monitor such weather parameters that can provide a more precise prediction of the PV power generated [13]. Therefore, the PV power plant integration into power systems must imply monitoring solutions. Moreover, Beránek et al. affirms that monitoring of PV system plants is an urgent and imperative activity for practical implementation of new ecologically clean solar plants [14].

Different solutions can be found in the specific literature to monitor PV power plants. Ramakrishna et al. affirm that the PV monitoring systems can be broadly classified as ground-based or space-based monitoring systems [15]. More specifically, some of these contributions are focused on monitoring locally PV data. In this way, Fuentes et al. describes a portable data logger based on standalone instruments [16]. LabVIEW has shown relevant characteristics for monitoring and communicating several devices simultaneously [17]. Bayrak et al. use a Labview data acquisition (DAQ) card for monitoring electrical measurement of a PV system [18]. Chouder et al. also present a detailed characterization of the performance and dynamic behavior of PV installations through LabVIEW real-time interface system [19]. Recently, a novel power line communication (PLC) method for a DC–DC power optimizer solution is proposed by Zhu et al. [20]. The data are modulated and then transmitted through the series-connected DC-power line to other DC–DC power optimizers. The parallel resonant coupling unit is used in [21] to monitor PV data into a high-frequency form to carry out the carrier communication. Wireless solutions to monitor PV installations at panel level have been also proposed by other authors. As an example, Ando et al. describes a complete wireless solution at panel level to estimate efficiency losses and anomalous aging of PV installations [22]. Similar contributions for individual monitoring of panels based on wireless technology can be found in [23]. An in–situ monitoring solutions for PV panels is proposed and evaluated by Papageorgas et al. in [24]. Moreno-García et al. presents an architecture of acquisition devices, including distributed wireless sensors, to monitor and supervise all the distributed devices in the plant [25]. An extension of this solution by detecting any failures or deviations in PV production can be found in [26]. A low-cost acquisition system to record data in micro SD card is presented by Fanourakis et al. in [27]. Regarding remote PV monitoring proposals, different contributions can be found in the specific literature. In this way, Zigbee technology has been proposed by different authors during recent years [28–31]. Li et al. also propose an on-line monitoring system based on Zigbee technology for Internet of Things purposes [32]. However, and according to [33], Zigbee technology is proven inefficient in large scale since it is not able to face up huge distances. A low cost IOT-based embedded system is described in [34]. This solution uses a GPRS module and a low cost microcontroller to send the power generated by a PV power plant. GSM voice channel for the communication of data has been also proposed, since the GSM network is readily available in rural areas [35]. As drawbacks, Pereira et al. affirm that this solution requires a SIM card with data transfer charging and can be installed only in places under phone coverage [36]. At residential level, an IoT solution based on Arduino with 3G connectivity technology is described and assessed in [37]. A comparison of different technologies—Ethernet, WiFi and ZigBee—for smart-house applications including RES is proposed in [38]. A user-friendly PV monitoring system based on a low-cost PLC is proposed by Han et al in [39]. A review focused on solutions for PV performance monitoring is discussed in [40].

From the Transmission and Distribution Network Operator point of view, a mass energy production coming from PV systems without the corresponding energy storage units and/or sufficient innovative electricity network architectures—such as micro-grids, smart-grids and web of cells—can cause severe disturbances [41]. In Europe, Mateo et al. contribute to overcome the barriers that hamper a large-scale integration of PV installations in the electricity distribution grids, being necessary the integration of advanced monitoring and operation systems [42]. As an additional example, in Germany, 90% of renewable system capacity is connected to distribution grids, and smart grid investments should be promoted by German DSOs [43]. Current power systems thus require modernization in terms of sensing, communication technologies, measurements, and automation technologies, and subsequently, smart power grids arise as a suitable solution [44,45].

Considering previous approaches, this work provides a step forward: a wireless low-cost open-source monitoring solution based on long-range (LoRa) technology able to communicate with remote PV power plants. The aim is thus to monitor in real time wide zones under study, covering long distances with minimum power consumption and maintenance. This study is in line with previous works of the authors focused on PV monitoring [46,47]; as well as in line with recent contributions where PLC and wireless are considered the best candidates for communication purposes [48], and wireless is poised to play a significant role in shaping the capabilities of future measurement systems [49]. Besides, the cost of open-source solutions is usually considerably lower than commercially available devices, with little loss of accuracy and precision [16]. Moreover, commercial solutions present some drawbacks, as can be found in recent PV monitoring system reviews [50]. The main contributions are summarized as follows:

- Wide areas, referred to remote PV installations, are controlled via and communicated through a low-cost open-source solution based on LoRa technology.
- Data are gathered from the PV installations in accordance with the current IEC-61724 standards and industrial, scientific, and medical (ISM) band use regulations.
- The proposed solution is flexible to exchange data in real time among PV power plants in terms of power generation and weather parameters.

The rest of the paper is structured as follows: Section 2 describes wireless sensor network technology and particularly the LoRa approach. Section 3 gives detailed information regarding our proposed solution. Section 4 offers extensive results, evaluating the performance of our solution. To this end, different testing processes are conducted by the authors. Finally, conclusions are discussed in Section 5.

2. Wireless Sensor Network: LoRa Solution

Wireless sensor network (WSN) is a mature field in technology to sense physical parameters and transmit them wirelessly out of the coverage range of the measurement in situ. This is a potential area of interest for the scientific community, reinforcing some beliefs about the necessity for further research initiatives in a new wireless-network paradigm. In this context, the low-power wide-area network (LPWAN) is a recent WSN-based technology that emerged as an alternative wireless monitoring solution [51]. Different applications and contributions can be found, mainly focused on the industrial environment. In particular, this technology is currently drawing much attention for managing assets over wide areas, such as the monitoring and control of PV power plants and the operation of distributed energy systems. The main characteristics are its excellent long-range, low-power consumption and reduced computation capacity (like a long-range WSN). In fact, the operation of a distributed energy system usually requires flexible and reliable communication systems. However, cable-based communications are, in many cases, an infeasible solution due to their complex installation and maintenance. In distributed energy systems with high penetration of renewables and small generation units connected at different voltage levels, this wireless technology can help to overcome the lack of information in the performance and generation of these installations [52].

LPWAN is a generic term that encompasses a group of technologies, allowing wide area communications at lower cost points and reduced power consumption. LPWAN technologies have arisen in both licensed and unlicensed markets, such as LTE-M, Sigfox, long range (LoRa), and narrow band (NB)-IoT. Among them, LoRa and NB-IoT are the most prominent, though they clearly present technical differences [53]. Presently, LoRa is an LPWAN approach receiving relevant consideration in the literature, because it can operate efficiently in unlicensed bands. Unlike LoRa, an NB-IoT network must be set up within an existing cellular network. This makes LoRa a more flexible solution than NB-IoT to meet the requirements of outlying districts [54]. It is worth noting that LoRa is inarguably the main actor in the current LPWAN scene, used in an unlicensed spectrum below 1 GHz and supported by many worldwide technology leaders (Cisco, Microchip, IBM, HP, etc.) [55]. From a technological point of view, LoRa provides a proprietary chirp spread spectrum (CSS) modulation to achieve communication distances greater than 700 km [56]. What makes LoRa stand out from other modulation methods is its unique spread spectrum technique, which provides robustness against interference and a very low minimum signal-to-noise ratio (SNR) for the receiver to be able to demodulate the signal. LoRa is thus a suitable solution for applications that require a very long battery lifetime and reduced cost. Moreover, as it strengthens, LoRa allows tuning of several physical transmission properties: the bandwidth and central frequency of the communication, the coding rate (CR, the ratio between the length of the packet and the length of the error-correction code), the transmission power, and the spreading factor (SF, defined as the ratio between the symbol rate and chip rate). Higher SF values enhance the sensitivity and range of communication at the expense of increasing the over-the-air time of the packets, thus consuming more transmission duty cycles (TDCs).

In the past few years, the interest of monitoring smart industries has increasingly become LoRa [57] as one technology solution demanded by many researchers [58,59]. Most of the contributions have been focused on analyzing the advantages, disadvantages, capabilities, and limits of current developments/deployments in several scenarios: industrial environments [60], civil infrastructures such as bridges [61] and public transport [62], line-of-sight and obstructed communications [63], and surveillance tasks to combat poachers in wildlife reserves in Africa [64]. In addition, LoRa performance has been compared to other LPWAN solutions such as Sigfox [65] and Weightless [66], as well as licensed options such as NB-IoT [67]. Other studies have dealt with the real scalability of current LoRa networks [68,69], the performance of their different configurations [70] or the download traffic analysis of these types of networks [71]. LoRa defines the physical level and LoRaWAN encompasses the link layer of the protocol stack and the system architecture [72]. LoRaWAN uses long-range star architecture in which gateways are used to relay messages between the end nodes and a central core network (see Figure 1). In a LoRaWAN scenario, nodes are not associated with a specific gateway. Instead, data transmitted by a node are typically received by multiple gateways. Furthermore, LoRaWAN uses the adaptive data rate (ADR) algorithm to estimate the CR and SF parameters under specific channel conditions. Subsequently, each gateway forwards the received packet from the end node to the cloud-based network server via standard IP connections. Different disadvantages can be identified when LoRaWAN solutions are implemented, intrinsic to operating in any ISM band. In particular, current international laws require a stringent duty cycle of 1%. This means the radio channel cannot be occupied more than 36 s per hour. In fact, this value is denoted as the maximum TDC allowed by the nodes to operate in ISM channels. This is an important concern for nodes managing critical assets (such as those found in the proposed solution), where LoRa and LoRaWAN must be able to report critical events within seconds. Therefore, node duty cycles should be set with the goal of reporting critical events under the entire conditions. It is precisely this type of situation that drew our attention and a question arose: 'Is it possible to obtain communications using LoRa technology, considering its stringent duty cycle under critical conditions (i.e., given the criticality of the reported event)?' Moreover, the end-node configuration is a crucial aspect for packet transmission purposes, since LoRa networks allow us to adjust not only frequency and power values, but also other parameters such as SF and CR, promoting robustness in the communications at the expense of increasing the packet over-the-air time.

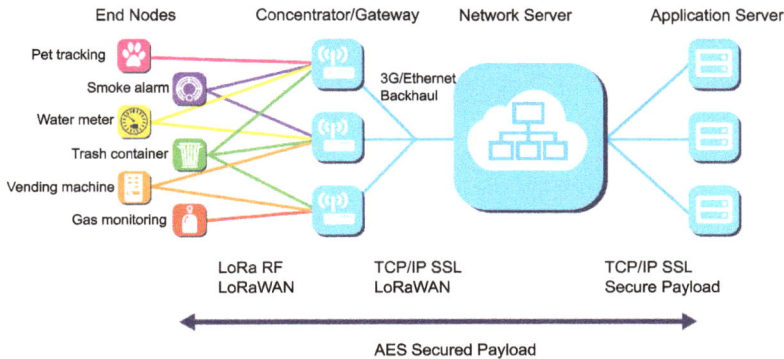

Figure 1. General overview of long-range (LoRa)/LoRa wireless area network (WAN) architecture.

To the best of the authors knowledge, contributions such as [73–75] use LoRa technology as the way of transmitting data collected by sensors in a PV Power Plant. In these contributions, authors installed a Lora communication module to dispatch packets to a Gateway including physical parameters such as current or temperature, without going into details about communication concerns. Unlike these works, we contribute with a solution which tackles these concerns more intensively than [73–75], ensuring the best performance for transmitting information from the PV installation to the Gateway and, consequently, increasing the system reliability. In this sense, and as will be discussed in Section 4, the transmission rate—bits per second—has been thoroughly tuned to achieve a suitable and acceptable RSSI (Received Signal Strength Indicator) and SNR figures in the proposed solution. For testing purposes, it has been considered the maximum SF allowed by the LoRa technology (SF12), with a CR value of 4/5 and below the maximum TDC.

3. Proposed Solution

3.1. General Overview: Topology

The proposed network architecture integrates LoRa–IoT infrastructure in a similar way to Figure 1. Therefore, by allowing for typical LoRa network topologies [76], a star topology including end devices, gateways, and a central network server is considered in the proposed solution. All selected sensors in charge of monitoring PV installations have to fulfill the IEC-61724 requirements. PV electrical data and weather parameters are gathered to estimate PV operating conditions and exchange meteorological and electrical data information. These data packets are sent to the LoRa gateway from the corresponding PV power plants. Subsequently, the received data are then forwarded to the network application via LAN connection. The design of the proposed system involves (i) end-node hardware selection, (ii) software-node configuration, and (iii) LoRa network server. These items are discussed in detail in the following.

3.2. End-Node Hardware: Sensors and Communication

The end node is in charge of gathering PV electrical data and weather conditions. Data packets are subsequently sent to the gateway. The main requirements accomplished by the end node are the following:

- Information from the PV module installed in the PV power plant is collected according to the current IEC-61724 standard requirements. These data provide relevant information for predictive maintenance purposes.
- Flexible, low-cost, and open-source solutions are required to carry out a suitable integration of the proposed system into real PV power plants.

- Nodes are able to run software, including a complete LoRa Class A. As mentioned, the end nodes operate under the license-exempt industrial, scientific, and medical (ISM) bands (EU 868 MHz/US 915 MHz) [77].

First, a hardware development platform was selected under the parameters of low cost and open-source software solutions. To this end, and taking into account previous experience, the Arduino platform [78], which offers a free development software environment to develop a prominent number of applications [79], was chosen. Other studies affirm the efficient reconfigurable security approach for WSN with Arduino-based systems [80]. Following these parameters, the Arduino One and Arduino Nano were considered as printed board platforms [81]. Both hardware solutions are based on ATmega328P with similar performance. As its main features, Arduino Uno comprises a 32 kB flash memory, 2 kB SRAM and 1 kB EEPROM, with 5 V operating voltage level and 14 digital I/O pins. The Arduino Nano is considered as a bridge between sensors and, for instance, a Raspberry Pi, which becomes it in a base station [82]. LoRa transceiver is also integrated in the device under the Arduino requirements [83].

Concerning the selected group of sensors, they have to be in line with the following requirements: they must (i) gather electrical PV parameters and weather conditions, and (ii) fulfill the IEC-61724 requirements. Moreover, the sensors are in accordance with previous works by the authors, where the same requirements were considered [16,84]. Local data collected by our proposed solution are thus able to estimate the PV module behavior and, in general, the PV installation performance. In terms of electrical data, AC and DC voltage and current variables are considered as parameters to be measured and collected for monitoring purposes. AC voltage measurement (V_{AC}) is implemented by an AC–AC power adaptor. An isolation transformer gives a physical separation and a quasi-sinusoidal waveform as an output signal. This signal is adapted by a voltage divider and sent to the Arduino board as an analog input. In a similar way, DC voltage (V_{DC}) is collected and adapted as an Arduino board analog input as well. For AC data (I_{AC}), a noninvasive Hall-effect sensor is provided for the proposed solution [85]. An accurate, low-offset, linear Hall sensor is selected and implemented by the authors to measure the DC current (I_{DC}) [86]. Both AC and DC sensors offer low-voltage output signals compatible with the Arduino input voltage range. With regard to weather parameters, the following variables are considered for monitoring purposes: solar irradiance, ambient temperature, and PV module temperature. To measure and gather solar irradiance, and assuming that the short-circuit current (ISC) is nearly proportional to the irradiance [87], it is measured in W/m^2 by a 5 Wp short-circuit encapsulated polycrystalline silicon module. A shunt resistance is chosen and implemented to adapt the voltage output within a suitable voltage range according to the Arduino analog input requirements. Calibration of this module was carried out by the authors through the CETENMA Solar TestBed, based on the global sunlight method available in [88]. Ambient temperature was measured near PV modules as an attempt to more accurately estimate the real environment of PV module conditions. A DHT22 temperature/humidity sensor with digital output was selected with this objective. The DHT22 sensor is directly supported by the Arduino IDE technology and, according to [89], it furnishes very accurate results with a fast refresh time. Other applications using the DHT22 sensor can be found in [90,91]. Most correlations in the scientific literature for PV electrical power as a function of the cell/module operating temperature and basic environmental variables are based on linear approaches [92]. Indeed, [93,94] affirm that PV module power output values depend linearly, but rather strongly, on the operating temperature. The authors note that the PV module temperature should be collected at the center of the back surface of the module and in the center of the array field location on the module, as pointed out by IEC-61829 method A [95]. In our proposed system, a low-cost solution employing the DS18S20 digital sensor is used with this goal. This digital thermometer achieves 9-bit Celsius temperature measurements, transferring them on a 1-wire bus [96]. A general diagram of the sensors and their connections with the Arduino board is depicted in Figure 2.

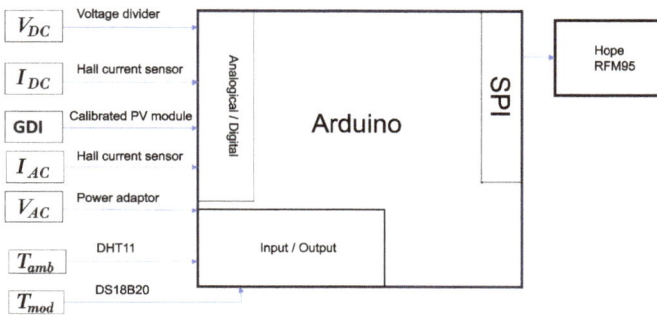

Figure 2. General diagram of sensors and Arduino connections.

The communication system is in charge of dispatching data packets thanks to a transceiver integrated into the Arduino board and operating in the EU-868 band [72]. The selected transceiver is the RFM95W module, fabricated by HOPE RF and configured as a LoRa TM modem [97]. In fact, the RFM95W configured for LoRa communication via 4-wire SPI bus was successfully tested in [98]. The main characteristics of this LoRa TM module are high efficiency and significant sensitivity (around −148 dBm). To achieve these advantages, this module is composed of a 6-GPIO interface configurable by software and with different interruptions usually linked to the operation of the RFM95W [99], and is able to support different modulations such as FSK/OOK, GFSK, MSK, or GMSK. Due to the small size of the RFM95W module, an adapter is required to provide breakout pins and the antenna plug-ins (Figure 3). Finally, the RFM95W works at 3.3 V and thus it cannot be directly connected to the Arduino Uno or Nano (both operating at 5 V). A voltage adapter is thus required to give the operating voltage range.

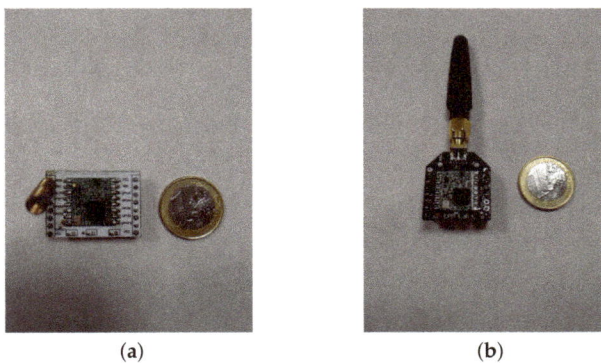

(a)   (b)

Figure 3. Dispatching data packets: implemented transceivers. (**a**) RFM95 with an ESP8266 module adapter. (**b**) HOPE RFM95W transceiver with breakout board and 868 MHz antenna.

3.3. End-Node Software Configuration

The end-node software configuration was subsequently conducted after selecting the end-node hardware solution along with the sensors. First, Arduino firmware was implemented and tested in order to obtain the PV power plant parameters and values under the IEC-61724 standard [100] and the ISM band-use regulations. The selected firmware was the nano-lmic-v1.51-F.ino, which was downloaded from the LoRa LMIC library [101]. The Arduino code follows activation-by-personalization (ABP) rules, available in the LMIC library [83]. SScripts to read the data format were also developed and the payload adjusted to contain full information gathered by the

Energies **2019**, *12*, 881

different sensors, following the format recommended by the IEC standard resolution. Please note that the payload has a length of 38 bytes, which has a relevant impact on the over-air time of the packets, as will be discussed in Section 4. The sampling period is another aspect to be considered in detail. In this sense, the IEC standard establishes that the sampling period of the different parameters under study varies proportionally to the solar irradiance. Under our operating conditions, parameters had to be sampled in intervals of 1 minute or less. On the other hand, in Europe, the duty cycle is regulated by the ETSI EN300.220 standard—Section 7.2.3—[102], which, as noted in Section 2, sets a duty cycle of 1%. Furthermore, from the spreadsheet developed and proposed by Matthijs Kooijman [103], the over-air time of the packets can be determined for different SFs. The over-air time corresponding to the worst case was then used to define our sampling rate. By considering our scenario, the duty cycle, and the over-air-time issue, data packets were sampled every 30 s. The parameters were then averaged and sent in time periods of 3 min. These values will be modified in a subsequent version to enforce the so-called *'Fair Access Policy'*, which limits the uplink airtime to 30 s per day and per node [104]. The downlink messages were set to 10 messages per day and per node.

3.4. LoRa Gateway/LoRa Network Server

A single-channel LoRa gateway was designed and implemented with a Raspberry Pi board. The LoRa radio communication module we selected is the Dragino Arduino shield from a Semtech SX1276 chip [105]. The Dragino shield can be directly connected to the Raspberry Pi. SPI connectors belonging to Dragino (MISO, MOSI, CLK, and NSSEL), VCC and GND are attached to the corresponding pins on the Raspberry Pi (CE0 on the RPI for SPI, *_nSSEL* and 3v3 for VCC). The operating system (OS) installed on the Raspberry Pi was Raspbian [106]. This single channel gateway software as well as its OS are supported by Thomas Telkamp at GitHub, and further information can be found in [107]. The single-channel LoRa gateway assembly and the coverage range are depicted in Figure 4. As described in Section 3.1, a LoRa network server is required to test the proposed application. With this aim, The Things Network (TTN) backend is used [108]. TTN is a community-driven project offering a free server to users; our proposed solution must connect the gateway to this free server. To this end, we created an account and registered the developed gateway. The gateway is located in Europe, selecting the European ISM band (868 MHz). After this setting process, the gateway is ready to be plugged in to the server. At this point, users can visualize the data packet transmission from the end nodes associated with this gateway, with a status of *'connected at web'*. In our case, two applications were developed: monitoring system testing and coverage range testing. Furthermore, the Arduino firmware implemented a different device address, 'DEVADDR', the network session key 'NWSKEY', and the application session key 'APPSKEY', which were developed and implemented before carrying out the corresponding validation tests discussed in detail in the following section.

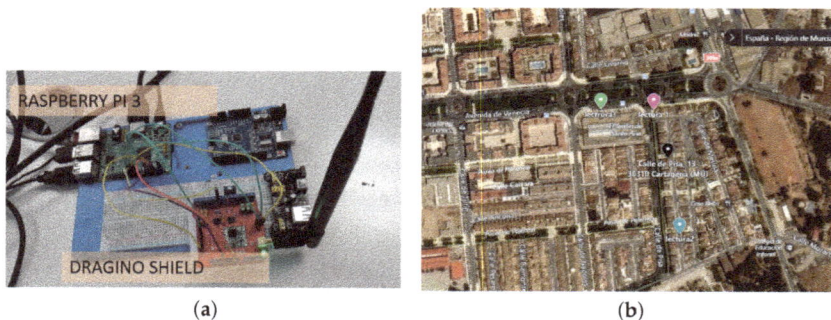

Figure 4. Single-channel gateway. (a) Gateway assembly. (b) Gateway coverage range.

3.5. Economic Evaluation: Cost-Effectiveness

Finally, and in terms of cost-effectiveness, the proposed monitoring system is in line with other contributions discussed in Sections 1 and 2 based on non-commercial solutions. Our system is flexible to be configured in different locations and PV installations. With the aim of offering a low cost system, the different hardware components are based on open-source projects with a high cost effectiveness threshold. Table 1 gives the monitoring node cost, which is lower than other commercial solutions as was previously discussed. Moreover, these commercial solutions usually provide less information and a reduced number of parameters.

Table 1. PV monitoring node cost.

Description	Number	Unit Price (Euro)	Total Price (Euro)
PV module (5 Wp, 22 V, 30 W)	1	8.75	8.75
Cement resistance 5 W 10 Ω 10 R 5%	1	0.13	0.13
AC-AC power supply adapter	1	4.95	4.95
Non-invasive AC–sensor	1	4.31	4.31
5 V DC-USB power adapter	1	2.40	2.40
2.54 mm PCB screw connector	6	0.17	1.02
Aluminium electrolytic capacitor 400 V	2	0.02	0.04
Metal film resistance 1 M 1.2 M 1.5 M 2 M 2.2 M Ω	6	0.05	0.30
Prototype PCB universal board	1	0.35	0.35
Outdoor enclosure and wiring	1	3.50	3.50
RFM95 LoRa Breakout + SMA connector antenna	1	5.95	5.95
HOPE RFM95	1	4.05	4.05
DS12820 temperature sensor	1	0.99	0.99
DHT temperature and humidity sensor	1	2.52	2.52
Total cost			39.26

4. Results

This section summarizes the test bed conducted to evaluate the proposed system. System assembly, coverage range, and performance evaluation of the PV power plant are described with the goal of adding value to our proposal in terms of a sensing, monitoring, and data packet transmission solution in an IoT scenario.

4.1. System Assembly

First, the selected hardware components and sensors (end nodes) were tested in a laboratory environment to evaluate their performance under controlled conditions. After this initial testing, components and sensors were connected and assembled to provide a feasible solution able to operate under real conditions. To facilitate the integration of the proposed monitoring system into real PV power plants, components and sensors corresponding to the end nodes were divided into two subnodes, including most connectors commonly used in real PV installations connected to the grid: (i) a principal subnode involving a main controller, a transceiver with the corresponding voltage level converter, and a set of batteries for power supply requirements (see Figure 5); (ii) a secondary subnode in charge of measuring and collecting PV module variables. To test the appropriateness of the global solution, these nodes were first deployed in the solar laboratory of CETENMA, located in the Industrial Park of Cartagena (southeast Spain). This facility includes measurement equipment to check the performance of PV power plants and modules. For testing purposes, a single 250 Wp monocrystalline PV module connected to an SF 250 W Soltec SolarFighter microinverter was used (see Figure 6).

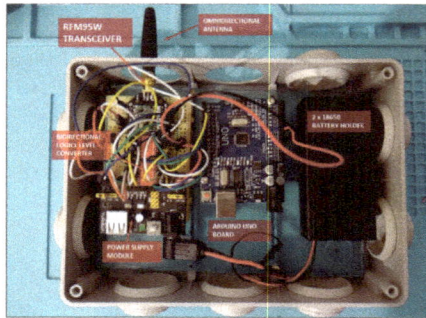

Figure 5. Detail of the system assembly. Principal subnode.

(a)   (b)

Figure 6. Testing system located at the CETENMA SolarLab (Spain). (**a**) SolarLab outdoor test site details. (**b**) Secondary box details.

4.2. Coverage Range Characterization

A relevant objective of this work was to evaluate the suitability of the proposed solution to measure, collect, and send data from PV installations to a remote gateway. The coverage range of the initial gateway, discussed in Section 3.3, was not enough to achieve the objectives searched in the test bed. Indeed, the location and low performance of the antenna used for this gateway were likely the biggest drawbacks of the initial poor coverage range. To overcome this limitation, another gateway was implemented and installed at the Universidad Politecnica de Cartagena (Cartagena, Spain). This additional Gateway is called the TM RG186 Series LoRa-enabled gateway [109]. It is located on the roof of a researching building on the university campus (see Figure 7). The gateway was registered with TTN as well.

To ensure the communication of the proposed system, we conducted a test of the signal power and coverage range. To this end, a Global Positioning System (GPS) module was integrated into the end node to transmit the location coordinates. To visualize the position of each end node, the TTN Mapper software tool [110] was installed in the network server. This additional tool provides further information, such as RSSI and SNR. Furthermore, each end node was configured with a transmission power of 14 dBm and SF12 to carry out the set of tests. During these tests, a new gateway was identified 15 km from our installation in the solar laboratory of CETENMA. Figure 8 illustrates these results corresponding to the coverage testing process.

Figure 7. TM RG186 series LoRa-enabled gateway (Universidad Politecnica de Cartagena, Spain).

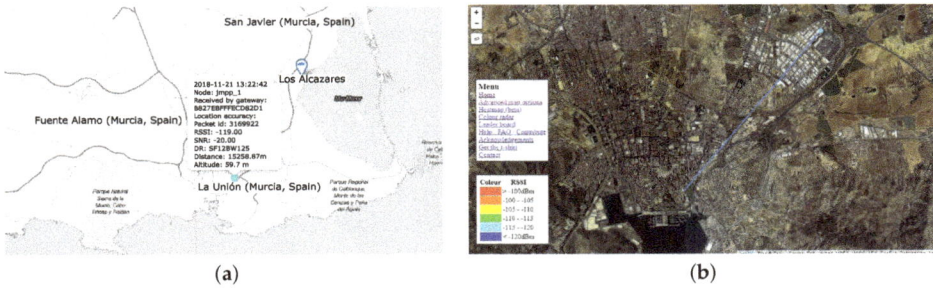

(a)   (b)

Figure 8. Coverage testing analysis. (**a**) Gateway identification map. (**b**) Coverage testing results on The Things Network (TTN) Mapper website.

4.3. PV Power Plant Performance Evaluation

Considering the different coverage testing processes we previously carried out, discussed in Section 4.2, a 5 kW PV installation connected to the grid and located on the university campus was deployed to assess the PV monitoring properties of the proposed solution (see Figure 9). Electrical and environmental data were gathered from the PV power plant and sent to the gateway to be evaluated and discussed in subsequent analysis. Additional parameters such as encrypted payload, received signal strength indicator (RSSI), air time, signal-to-noise ratio (SNR), number of packets, and channel included in the LoRa packet are also available and can be downloaded from the TTN website.

Figure 9. End nodes of 5 kWp PV installation monitoring.

A packet size of 38 bytes is considered enough to cover all the parameters for PV power plant monitoring purposes. The transmission power is 14 dBm and the SF metric is tuned from 10 to 12, which influences the data packet over-air time. Figure 10 shows the theoretical time on-air (ms) for each SF configuration depending on the payload length. These results allowed us to configure the sampling period for each SF: 60, 120, and 180 s for SF10, SF11, and SF12, respectively. Our test bed was conducted for 24 h. This time interval was enough to evaluate each SF configuration. It is relevant to point out that SF10 involves no reception of packets in the gateway, as a consequence of different concerns, such as (i) the limited over-air time due to the distance between the device and the gateway, around 4 km; (ii) the locations of end nodes; and (iii) the conditions of the signal propagation.

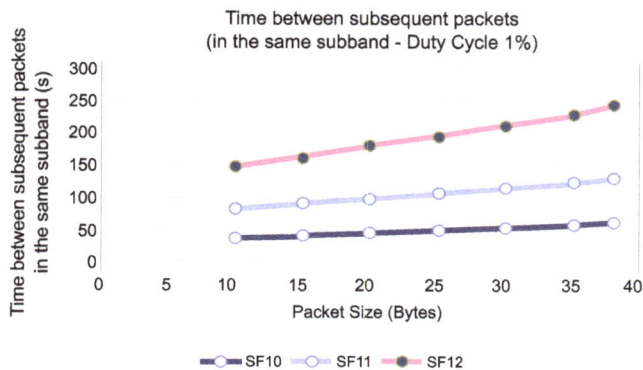

Figure 10. Theoretical time-on-air (ms) for different tested SF configurations.

Figure 11 illustrates the RSSI and SNR results obtained in November 2018 for SF11 and SF12. Table 2 shows additional metrics also discussed in this work: packet delivery ratio and time intervals between different data packets (inter-arrival time). The packet delivery ratio is defined as the ratio between the packets successfully received and the total data packets sent by the end nodes.

The inter-arrival time is determined by the time interval value corresponding to each packet received by the TTN web application. For SF11, our study reveals low RSSI and SNR values. As an example, the minimum signal strength in WiFi technology provides basic connectivity with reliable packet delivery around $-80/-90$ dBm. Concerning SF12, the RSSI and SNR metrics improved around 10%. However, the time interval between packets increased to 115 s, which is reasonable due to greater over-air time of the packets. In terms of the packet delivery ratio, SF11 showed poor performance, with most of the packets corrupted or completely lost. To overcome this drawback, it is necessary to increase the spreading factor (SF), which allows us to improve the metric sharply. It is remarkable that these outcomes are in line with recent contributions [111–113], which corroborates one of main advantages of using LoRa: its sensibility. In this respect, weak signals can stimulate the electronic communication of the LoRa device, resulting in successful packet delivery to the gateway. Finally, to verify that the payload is being decrypted correctly, data received by the TTN application are compared to the same data collected by a data logger of the test stand (see Figure 12). These results validate the feasibility and reliability of our proposal, as well as the accuracy of the implemented monitoring and communication solution.

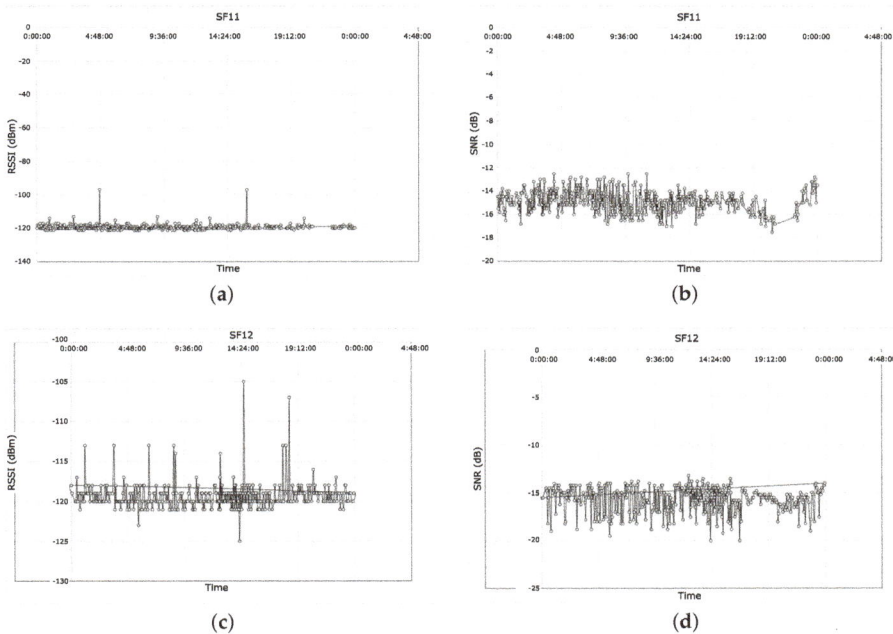

Figure 11. Coverage range tests. (**a**) Received signal strength indicator (RSSI) data (SF11). (**b**) Signal-to-noise ratio (SNR) results (SF11). (**c**) RSSI data (SF12). (**d**) SNR results (SF12).

Table 2. Packet delivery rate and inter-arrival time metrics. SF, spreading factor.

SF	Packet Data Sent	Packet Data Received	Packet Delivery Rate	Average Time between Packets (s)	Inter-Arrival Time (s)
11	559	307	55%	190.85	131.00
12	402	364	91%	273.13	246.00

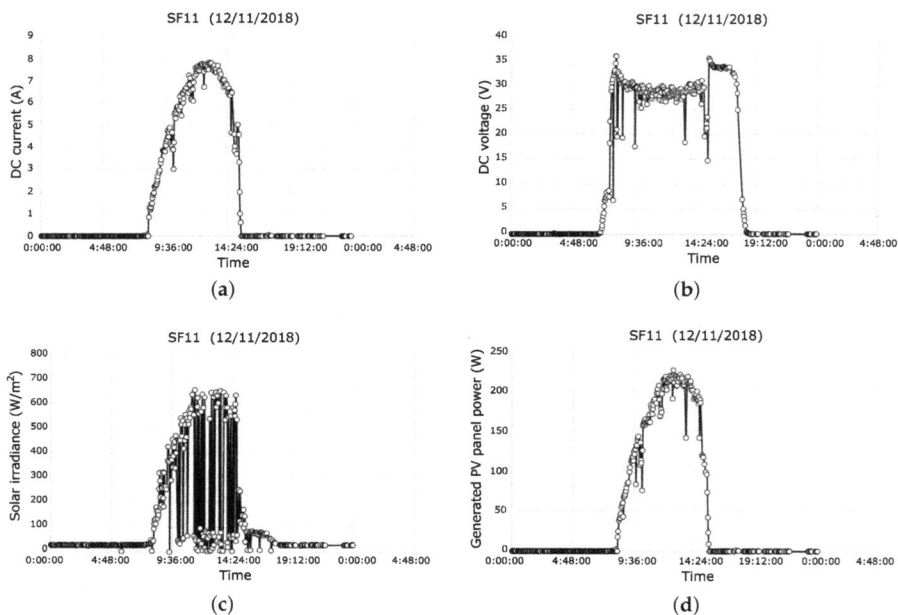

Figure 12. PV example of collected data (12 November 2018). (**a**) DC current data. (**b**) DC voltage data. (**c**) Solar irradiance data. (**d**) Generated active power.

5. Conclusions

A low-cost, open-source solution to monitor PV power plants was designed and evaluated. Our alternative system provides a powerful and straightforward solution, facilitating the integration of this renewable energy source into current power systems. As a novelty, long-range communication technology denoted as LoRa enables data transmission to a remote gateway, which allows us to evaluate the PV installation performance in real time. Extensive electrical and meteorological information is also available from the monitoring system. These data can be applied for predictive maintenance purposes. Moreover, these data have a remarkable impact on grid reliability and PV forecast accuracy.

A PV power plant connected to the grid (5 kW rate power) and located on the university campus (southeast Spain) was used to evaluate the monitoring system. Different field-test campaigns were conducted by the authors. From the results, it can be affirmed that aspects such as the distance between source and destination, the line-of-sight between source and destination, and propagation issues have a clear influence on the appropriate data reception process. Our study demonstrates that scenarios with a high spreading factor value (SF11 and SF12) satisfy an accurate reception of data packets. However, the corresponding over-air time considerably limits the number of transmissions. To overcome this drawback, the sampling time was adjusted in line with the packet air-time and according to the SF value. One of the main limitations of the LoRa solution is its restricted duty cycle (1%), which was taken into account for the testing process. Received data were also compared to data-logger equipment connected in situ in the PV installation. This comparison validates the electrical and meteorological variables gathered by different sensors, resulting in errors lower than 0.5%. Our proposed solution thus offers an alternative system to be implemented in remote PV power plants with the goal of monitoring and dispatching electrical and meteorological data.

Author Contributions: Data curation, J.M.P.-P.; Formal analysis, A.M.-G. and A.J.G.-S.; Funding acquisition, A.M.-A.; Investigation, A.J.G.-S.; Resources, J.M.P.-P.; Supervision, A.M.-G.; Visualization, A.M.-A.; Writing—original draft, A.M.-G.; Writing—review and editing, A.M.-G.

Funding: This work was partially supported by the Spanish agreement (2017) between the Institute for Development of the Region of Murcia (INFO) and the Technological Center for Energy and Environment (CETENMA). The paper includes results of activities conducted under the Research Program for Groups of Scientific Excellence at Region of Murcia (Spain), the Seneca Foundation, and the Agency for Science and Technology of the Region of Murcia (Spain). This work was also supported by project AIM, Ref. TEC2016-76465-C2-1-R (AEI/FEDER, UE).

Acknowledgments: The authors thank the staff of the Universidad Politécnica de Cartagena (Spain) for services and facilities provided.

Conflicts of Interest: The authors declare no conflict of interest.

Abbreviations

The following abbreviations are used in this manuscript:

ABP	Activation-by-personalisation
ADR	Adaptive data rate
CR	Coding rate
CSS	Chirp spread spectrum
ISC	Short-circuit current
LPWAN	Low-power wide-area networks
RSSI	Received signal strength indicator
SARIMA	Seasonal autoregressive integrated moving average
SF	Spreading factor
SNR	Signal-to-noise ratio
TDC	Transmission duty cycle
TTN	The Things Networo
WSN	Wireless sensor network

References

1. Panwar, N.; Kaushik, S.; Kothari, S. Role of renewable energy sources in environmental protection: A review. *Renew. Sustain. Energy Rev.* **2011**, *15*, 1513–1524. [CrossRef]
2. Verzijlbergh, R.; Vries, L.D.; Dijkema, G.; Herder, P. Institutional challenges caused by the integration of renewable energy sources in the European electricity sector. *Renew. Sustain. Energy Rev.* **2017**, *75*, 660–667. [CrossRef]
3. Nehrir, M.H.; Wang, C.; Strunz, K.; Aki, H.; Ramakumar, R.; Bing, J.; Miao, Z.; Salameh, Z. A Review of Hybrid Renewable/Alternative Energy Systems for Electric Power Generation: Configurations, Control, and Applications. *IEEE Trans. Sustain. Energy* **2011**, *2*, 392–403. [CrossRef]
4. Mancarella, P.; Panteli, M. Influence of Extreme Weather and Climate Change on the Resilience of Power Systems: Impacts and Possible Mitigation Strategies. *Electr. Power Syst. Res.* **2015**, *127*, 259–270.
5. Brouwer, A.S.; van den Broek, M.; Seebregts, A.; Faaij, A. Impacts of large-scale Intermittent Renewable Energy Sources on electricity systems, and how these can be modeled. *Renew. Sustain. Energy Rev.* **2014**, *33*, 443–466. [CrossRef]
6. Liang, X.; Mazin, H.E.; Reza, S.E. Probabilistic generation and transmission planning with renewable energy integration. In Proceedings of the 2017 IEEE/IAS 53rd Industrial and Commercial Power Systems Technical Conference (ICPS), Niagara Falls, ON, Canada, 6–11 May 2017; pp. 1–9. [CrossRef]
7. Kamaruzzaman, Z. Effect of grid-connected photovoltaic systems on static and dynamic voltage stability with analysis techniques—A review. *Przegląd Elektrotechniczny* **2015**, *1*, 136–140. [CrossRef]
8. Hung, D.Q.; Shah, M.R.; Mithulananthan, N. Technical Challenges, Security and Risk in Grid Integration of Renewable Energy. In *Smart Power Systems and Renewable Energy System Integration*; Jayaweera, D., Ed.; Springer International Publishing: Cham, Switzerland, 2016; pp. 99–118. [CrossRef]
9. Chiandone, M.; Sulligoi, G.; Massucco, S.; Silvestro, F. Hierarchical Voltage Regulation of Transmission Systems with Renewable Power Plants: An overview of the Italian case. In Proceedings of the 3rd Renewable Power Generation Conference, Naples, Italy, 24–25 September 2014. [CrossRef]
10. Behera, M.K.; Majumder, I.; Nayak, N. Solar photovoltaic power forecasting using optimized modified extreme learning machine technique. *Eng. Sci. Technol. Int. J.* **2018**, *21*, 428–438. [CrossRef]

11. Wang, F.; Zhen, Z.; Liu, C.; Mi, Z.; Hodge, B.M.; Shafie-khah, M.; Catalão, J.P. Image phase shift invariance based cloud motion displacement vector calculation method for ultra-short-term solar PV power forecasting. *Energy Convers. Manag.* **2018**, *157*, 123–135. [CrossRef]

12. Petinrin, J.; Shaaban, M. Impact of renewable generation on voltage control in distribution systems. *Renew. Sustain. Energy Rev.* **2016**, *65*, 770–783. [CrossRef]

13. Wolff, B.; Lorenz, E.; Kramer, O. Statistical Learning for Short-Term Photovoltaic Power Predictions. In *Studies in Computational Intelligence Computational Sustainability*; Springer: Berlin, Germany, 2016; Volume 645.

14. Beránek, V.; Olsan, T.; Libra, M.; Poulek, V.; Sedlacek, J.; Dang, M.Q.; Tyukhov, I. New Monitoring System for Photovoltaic Power Plants' Management. *Energies* **2018**, *11*, 2495. [CrossRef]

15. Madeti, S.R.; Singh, S. Monitoring system for photovoltaic plants: A review. *Renew. Sustain. Energy Rev.* **2017**, *67*, 1180–1207. [CrossRef]

16. Fuentes, M.; Vivar, M.; Burgos, J.; Aguilera, J.; Vacas, J. Design of an accurate, low-cost autonomous data logger for PV system monitoring using Arduino^{TM} that complies with IEC standards. *Sol. Energy Mater. Sol. Cells* **2014**, *130*, 529–543. [CrossRef]

17. Belhadj Ahmed, C.; Kassas, M.; Essamuddin Ahmed, S. PV-standalone monitoring system performance using LabVIEW. *Int. J. Smart Grid Clean Energy* **2014**, *3*, 44–50. [CrossRef]

18. Bayarak, G.; Cebeci, M. Monitoring a grid connected PV power generation system with labview. In Proceedings of the 2013 International Conference on Renewable Energy Research and Applications (ICRERA), Madrid, Spain, 20–23 October 2013; pp. 562–567. [CrossRef]

19. Chouder, A.; Silvestre, S.; Taghezouit, B.; Karatepe, E. Monitoring, modelling and simulation of PV systems using LabVIEW. *Sol. Energy* **2013**, *91*, 337–349. [CrossRef]

20. Zhu, Y.; Wu, J.; Wang, R.; Lin, Z.; He, X. Embedding Power Line Communication in Photovoltaic Optimizer by Modulating Data in Power Control Loop. *IEEE Trans. Ind. Electron.* **2019**, *66*, 3948–3958. [CrossRef]

21. Mao, W.; Zhang, X.; Cao, R.; Wang, F.; Zhao, T.; Xu, L. A Research on Power Line Communication Based on Parallel Resonant Coupling Technology in PV Module Monitoring. *IEEE Trans. Ind. Electron.* **2018**, *65*, 2653–2662. [CrossRef]

22. Ando, B.; Baglio, S.; Pistorio, A.; Tina, G.; Ventura, C. Sentinella: Smart Monitoring of Photovoltaic Systems at Panel Level. *IEEE Trans. Instrum. Meas.* **2015**, *64*, 2188–2199. [CrossRef]

23. Prieto, M.; Pernía, A.; Nuno, F.; Diaz, J.; Villegas, P. Development of a Wireless Sensor Network for Individual Monitoring of Panels in a Photovoltaic Plant. *Sensors* **2014**, *14*, 2379–2396. [CrossRef] [PubMed]

24. Papageorgas, P.; Piromalis, D.; Antonakoglou, K.; Vokas, G.; Tseles, D.; Arvanitis, K. Smart Solar Panels: In-situ Monitoring of Photovoltaic Panels based on Wired and Wireless Sensor Networks. *Energy Procedia* **2013**, *36*, 535–545. [CrossRef]

25. Moreno-García, I.; Palacios-García, E.; Pallares-López, V.; Santiago, I.; González-Redondo, M.; Varo-Martínez, M.; Real-Calvo, R. Real–Time Monitoring System for a Utility–Scale Photovoltaic Power Plant. *Sensors* **2016**, *16*, 770. [CrossRef]

26. Moreno-García, I.M.; Palacios-García, E.J.; Santiago, I.; Pallares-López, V.; Moreno-Munoz, A. Performance monitoring of a solar photovoltaic power plant using an advanced real-time system. In Proceedings of the 2016 IEEE 16th International Conference on Environment and Electrical Engineering (EEEIC), Florence, Italy, 7–10 June 2016; pp. 1–6. [CrossRef]

27. Fanourakis, S.; Wang, K.; McCarthy, P.; Jiao, L. Low-cost data acquisition systems for photovoltaic system monitoring and usage statistics. *Earth Environ. Sci.* **2017**, *93*, 012048. [CrossRef]

28. Katsioulis, V.; Karapidakis, E.; Hadjinicolaou, M.; Tsikalakis, A. Wireless Monitoring and Remote Control of PV Systems Based on the ZigBee Protocol. In *Technological Innovation for Sustainability*; Camarinha-Matos, L.M., Ed.; Springer: Berlin/Heidelberg, Germany, 2011; pp. 297–304.

29. Ben Belghith, O.; Lassaad, S. Remote GSM module monitoring and Photovoltaic system control. In Proceedings of the 2014 First International Conference on Green Energy ICGE, Sfax, Tunisia, 25–27 March 2014; pp. 188–192. [CrossRef]

30. Zahurul, S.; Mariun, N.; Kah, L.; Hizam, H.; Othman, M.L.; Abidin, I.Z.; Norman, Y. A novel Zigbee-based data acquisition system for distributed photovoltaic generation in smart grid. In Proceedings of the 2015 IEEE Innovative Smart Grid Technologies—Asia (ISGT ASIA), Chengdu, China, 21–24 May 2015; pp. 1–6. [CrossRef]

31. Shariff, F.; Rahim, N.A.; Hew, W.P. Zigbee-based data acquisition system for online monitoring of grid-connected photovoltaic system. *Expert Syst. Appl.* **2015**, *42*, 1730–1742. [CrossRef]

32. Li, Y.F.; Lin, P.J.; Zhou, H.F.; Chen, Z.C.; Wu, L.J.; Cheng, S.Y.; Su, F.P. On-line monitoring system of PV array based on internet of things technology. *Earth Environ. Sci.* **2017**, *93*, 012078. [CrossRef]

33. Adhya, S.; Saha, D.; Das, A.; Jana, J.; Saha, H. An IoT based smart solar photovoltaic remote monitoring and control unit. In Proceedings of the 2016 2nd International Conference on Control, Instrumentation, Energy Communication (CIEC), Kolkata, India, 28–30 January 2016; pp. 432–436. [CrossRef]

34. Kekre, A.; Gawre, S.K. Solar photovoltaic remote monitoring system using IOT. In Proceedings of the 2017 International Conference on Recent Innovations in Signal processing and Embedded Systems (RISE), Bhopal, India, 27–29 October 2017; pp. 619–623. [CrossRef]

35. Tejwani, R.; Kumar, G.; Solanki, C. Remote Monitoring for Solar Photovoltaic Systems in Rural Application Using GSM Voice Channel. *Energy Procedia* **2014**, *57*, 1526–1535. [CrossRef]

36. Pereira, R.I.; Dupont, I.M.; Carvalho, P.C.; Jucá, S.C. IoT embedded linux system based on Raspberry Pi applied to real-time cloud monitoring of a decentralized photovoltaic plant. *Measurement* **2018**, *114*, 286–297. [CrossRef]

37. López-Vargas, A.; Fuentes, M.; Vivar, M. IoT Application for Real-Time Monitoring of Solar Home Systems Based on Arduino™ With 3G Connectivity. *IEEE Sens. J.* **2019**, *19*, 679–691. [CrossRef]

38. Ahmed, M.A.; Kang, Y.C.; Kim, Y.C. Communication Network Architectures for Smart-House with Renewable Energy Resources. *Energies* **2015**, *8*, 8716–8735. [CrossRef]

39. Han, J.; Lee, I.; Kim, S. User-friendly monitoring system for residential PV system based on low-cost power line communication. *IEEE Trans. Consum. Electron.* **2015**, *61*, 175–180. [CrossRef]

40. Hadjipanayi, M.; Koumparou, I.; Philippou, N.; Paraskeva, V.; Phinikarides, A.; Makrides, G.; Efthymiou, V.; Georghiou, G. Prospects of photovoltaics in southern European, Mediterranean and Middle East regions. *Renew. Energy* **2016**, *92*, 58–74. [CrossRef]

41. Kyritsis, A.; Voglitsis, D.; Papanikolaou, N.; Tselepis, S.; Christodoulou, C.; Gonos, I.; Kalogirou, S. Evolution of PV systems in Greece and review of applicable solutions for higher penetration levels. *Renew. Energy* **2017**, *109*, 487–499. [CrossRef]

42. Mateo, C.; Frías, P.; Cossent, R.; Sonvilla, P.; Barth, B. Overcoming the barriers that hamper a large-scale integration of solar photovoltaic power generation in European distribution grids. *Sol. Energy* **2017**, *153*, 574–583. [CrossRef]

43. Matschoss, P.; Bayer, B.; Thomas, H.; Marian, A. The German incentive regulation and its practical impact on the grid integration of renewable energy systems. *Renew. Energy* **2019**, *134*, 727–738. [CrossRef]

44. Colak, I.; Sagiroglu, S.; Fulli, G.; Yesilbudak, M.; Covrig, C.F. A survey on the critical issues in smart grid technologies. *Renew. Sustain. Energy Rev.* **2016**, *54*, 396–405. [CrossRef]

45. Hancke, G.P.; Silva, B.; Hancke, G.P., Jr. The Role of Advanced Sensing in Smart Cities. *Sensors* **2013**, *13*, 393–425. [CrossRef] [PubMed]

46. Molina-Garcia, A.; Campelo, J.C.; Blanc, S.; Serrano, J.J.; Garcia-Sanchez, T.; Bueso, M.C. A Decentralized Wireless Solution to Monitor and Diagnose PV Solar Module Performance Based on Symmetrized-Shifted Gompertz Functions. *Sensors* **2015**, *15*, 18459–18479. [CrossRef] [PubMed]

47. Paredes-Parra, J.M.; Mateo-Aroca, A.; Silvente-Niñirola, G.; Bueso, M.C.; Molina-Garcia, A. PV Module Monitoring System Based on Low-Cost Solutions: Wireless Raspberry Application and Assessment. *Energies* **2018**, *11*. [CrossRef]

48. Madeti, S.; Singh, S. Comparative analysis of solar photovoltaic monitoring systems. *AIP Conf. Proc.* **2017**, *1859*, 1–6. [CrossRef]

49. Zahran, M.; Atia, Y.; Alhosseen, A.; El-Sayed, I. Wired and wireless remote control of PV system. *WSEAS Trans. Syst. Control* **2010**, *5*, 656–666.

50. Rahman, M.M.; Selvaraj, J.; Rahim, N.; Hasanuzzaman, M. Global modern monitoring systems for PV based power generation: A review. *Renew. Sustain. Energy Rev.* **2018**, *82*, 4142–4158. [CrossRef]

51. Usman, R.; Parag, K.; Mahesh, S. Low Power Wide Area Networks: An Overview. *IEEE Commun. Surv. Tutor.* **2017**, *19*, 855–873.

52. Song, Y.; Lin, J.; Tang, M.; Dong, S. An Internet of Energy Things Based on Wireless LPWAN. *Engineering* **2017**, *3*, 460–466. [CrossRef]

53. Sinha, R.S.; Wei, Y.; Hwang, S.H. A survey on LPWA technology: LoRa and NB-IoT. *ICT Express* **2017**, *3*, 14–21. [CrossRef]

54. Migabo, E.; Djouani, K.; Kurien, A.; Olwal, T. A Comparative Survey Study on LPWA Networks: LoRa and NB–IoT. In Proceedings of the Future Technologies Conference (FTC), Vancouver, BC, Canada, 29–30 November 2017; pp. 1045–1051.

55. LoRa-Alliance. The Internet of Things—An explosion of Connected Possibility. Available online: https://docs.wixstatic.com/ugd/eccc1ade5fda268ed945e885a43a39b38752 (accessed on 28 November 2018).

56. LoRa-Alliance. The Internet of Things—An Explosion of Connected Possibility. LoRa-Alliance. Available online: https://www.thethingsnetwork.org/article/ground-breaking-world-record-lorawan-packet-received-at-702-km-436-miles-distance (accessed on 28 November 2018).

57. LoRa. LoRa-Alliance. 2018. Available online: https://lora-alliance.org (accessed on 28 November 2018).

58. Magrin, D.; Centenaro, M.; Vangelista, L. Performance evaluation of LoRa networks in a smart city scenario. In Proceedings of the IEEE International Conference on Communications (ICC), Kansas , MO, USA, 20–24 May 2017; pp. 1–7.

59. Sorensen, R.; Kim, D.; Nielsen, J.; Popovski, P. Analysis of Latency and MAC–layer Performance for Class A LoRaWAN. *IEEE Wirel. Commun. Lett.* **2017**, *6*, 566–569. [CrossRef]

60. Haxhibeqiri, J.; Karaagac, A.; den Abeele, F.V.; Joseph, W.; Moerman, I.; Hoebeke, J. LoRa indoor coverage and performance in an industrial environment: Case study. In Proceedings of the 22nd IEEE International Conference on Emerging Technologies and Factory Automation (ETFA), Limassol, Cyprus, 12–15 September 2017; pp. 1–8.

61. Orfei, F.; Mezzetti, C.B.; Cottone, F. Vibrations powered LoRa sensor: An electromechanical energy harvester working on a real bridge. In Proceedings of the 2016 IEEE Sensors, Orlando, FL, USA, 30 October–3 November 2016; pp. 1–3.

62. James, J.G.; Nair, S. Efficient, real–time tracking of public transport, using LoRaWAN and RF transceivers. In Proceedings of the TENCON 2017 IEEE Region 10 Conference, Penang, Malaysia, 5–8 November 2017; pp. 2258–2261.

63. Rahman, A.; Suryanegara, M. The development of IoT LoRa: A performance evaluation on LoS and Non-LoS environment at 915 MHz ISM frequency. In Proceedings of the International Conference on Signals and Systems (ICSigSys), Bali, Indonesia, 16–18 May 2017; pp. 163–167.

64. Rwanda Parks. Available online: https://www.theverge.com/2017/7/20/16002752/smart-park-rwanda-akagera-poaching-lorawan (accessed on 28 November 2018).

65. Vejlgaard, B.; Lauridsen, M.; Nguyen, H.; Kovacs, I.; Mogensen, P.; Sorensen, M. Coverage and Capacity Analysis of Sigfox, LoRa, GPRS, and NB-IoT. In Proceedings of the IEEE 85th Vehicular Technology Conference (VTC Spring), Sydney, Australia, 4–7 June 2017; pp. 1–5.

66. Goursaud, C.; Gorce, J. Dedicated networks for IoT: PHY–MAC state of the art and challenges. *EAI Endorsed Transac. Int. Things* **2015**. . [CrossRef]

67. Persia, S.; Carciofi, C.; Faccioli, M. NB-IoT and LoRA connectivity analysis for M2M/IoT smart grids applications. In Proceedings of the AEIT International Annual Conference, Cagliari, Italy, 20–22 September 2017; pp. 1–6.

68. Martin, B.; Utz, R.; Thiemo, V.; Juan, M.A. Do LoRa Low-Power Wide-Area Networks Scale? In Proceedings of the 19th ACM International Conference on Modeling, Analysis and Simulation of Wireless and Mobile Systems, Malta, 13–17 November 2016; pp. 59–67. [CrossRef]

69. Georgiou, O.; Raza, U. Low Power Wide Area Network Analysis: Can LoRa Scale? *IEEE Wirel. Commun. Lett.* **2017**, *6*, 162–165. [CrossRef]

70. Bor, M.; Roedig, U. LoRa Transmission Parameter Selection. In Proceedings of the 2017 13th International Conference on Distributed Computing in Sensor Systems (DCOSS), Ottawa, ON, Canada, 5–7 June 2017. [CrossRef]

71. Alexandru-Ioan, P.; Usman, R.; Parag, K.; Mahesh, S. Does Bidirectional Traffic Do More Harm Than Good in LoRaWAN Based LPWA Networks? In Proceedings of the GLOBECOM 2017 IEEE Global Communications Conference, Singapore, 4–8 December 2017. [CrossRef]

72. Alliance, L. *What Is LoRaWAN*; A Technical Overview; LoRaWAN: London, UK, 2015.

73. Kraemer, F.; Ammar, D.; Bråten, A.; Tamkittikhun, N.; Palma, D. Solar Energy Prediction for Constrained IoT Nodes Based on Public Weather Forecasts. In Proceedings of the Seventh International Conference on the Internet of Things, Linz, Austria, 22–25 October 2017. [CrossRef]

74. Shuda, J.; Rix, A.; Booysen, M.T. Towards Module-Level Performance and Health Monitoring of Solar PV Plants Using LoRa Wireless Sensor Networks. In Proceedings of the 2018 IEEE PES/IAS Power Africa, Cape Town, South Africa, 28–29 June 2018. [CrossRef]

75. Choi, C.; Jeong, J.; Lee, I.; Park, W. LoRa based renewable energy monitoring system with open IoT platform. In Proceedings of the 2018 International Conference on Electronics, Information, and Communication (ICEIC), Honolulu, HI, USA, 24–27 January 2018; pp. 1–2. [CrossRef]

76. De Carvalho Silva, J.; Rodrigues, J.; Alberti, A.; Solic, P.; Aquino, A. LoRaWAN—A Low Power WAN Protocol for Internet of Things: A Review and Opportunities. In Proceedings of the 2017 2nd International Multidisciplinary Conference on Computer and Energy Science (SpliTech), Split, Croatia, 12–14 July 2017.

77. Nolan, K.E.; Guibene, W.; Kelly, M.Y. An evaluation of low power wide area network technologies for the Internet of Things. In Proceedings of the 2016 International Wireless Communications and Mobile Computing Conference (IWCMC), Paphos, Cyprus, 5–9 September 2016; pp. 439–444. [CrossRef]

78. Arduino. Arduino.org. Available online: http://arduino.cc/ (accessed on 28 November 2018).

79. Galadima, A.A. Arduino as a learning tool. In Proceedings of the 2014 11th International Conference on Electronics, Computer and Computation (ICECCO), Abuja, Nigeria, 29 September–1 October 2014; pp. 1–4. [CrossRef]

80. Kabir, A.F.M.S.; Shorif, M.A.; Li, H.; Yu, Q. A study of secured wireless sensor networks with XBee and Arduino. In Proceedings of the 2014 2nd International Conference on Systems and Informatics (ICSAI 2014), Shanghai, China, 15–17 November 2014; pp. 492–496. [CrossRef]

81. Arduino Manual. Arduino Microcontroller. Arduino.org. Available online: http://arduino.cc/en/Reference/HomePage (accessed on 28 November 2018).

82. Deshmukh, A.D.; Shinde, U.B. A low cost environment monitoring system using raspberry Pi and arduino with Zigbee. In Proceedings of the 2016 International Conference on Inventive Computation Technologies (ICICT), Coimbatore, India, 26–27 August 2016; Volume 3, pp. 1–6. [CrossRef]

83. Telkamp, T. IBM LMIC v1.5 (LoRaWAN in C) Adapted to Run under the Arduino Environment. Technical Report. 2016. Available online: https://github.com/things-nyc/testnode-arm-cortex-mbed-lmic-1.5 (accessed on 28 November 2018).

84. Han, J.; Jeong, J.D.; Lee, I.; Kim, S.H. Low-cost monitoring of photovoltaic systems at panel level in residential homes based on power line communication. *IEEE Trans. Consum. Electron.* **2017**, *63*, 435–441. [CrossRef]

85. YHDC. STC-013-000 Datasheet. Available online: https://datasheet4u.com/datasheet-pdf/YHDC/STC-013-000/pdf.php?id=1089416 (accessed on 28 November 2018).

86. MicroSystems, A. ACS712–Datasheet. Available online: http://www.alldatasheet.com/datasheet-pdf/pdf/168326/ALLEGRO/ACS712.html (accessed on 28 November 2018).

87. Vicente, E.M.; Moreno, R.L.; Ribeiro, E.R. MPPT Technique Based on Current and Temperature Measurements. *Int. J. Photoenergy* **2015**, 1–9. [CrossRef]

88. Harald, M.; Willem, Z.; Ewan, D.; Heinz, O. Calibration of photovoltaic reference cells by global sunlight method. *Metrologia* **2005**, *42*, 360. [CrossRef]

89. Priya, C.G.; AbishekPandu, M.; Chandra, B. Automatic plant monitoring and controlling system over GSM using sensors. In Proceedings of the 2017 IEEE Technological Innovations in ICT for Agriculture and Rural Development (TIAR), Chennai, India, 7–8 April 2017; pp. 173–176. [CrossRef]

90. Hulea, M.; Mois, G.; Folea, S.; Miclea, L.; Biscu, V. Wi-sensors: A low power Wi-Fi solution for temperature and humidity measurement. In Proceedings of the 39th Annual Conference of the IEEE Industrial Electronics Society, Vienna, Austria, 10–13 November 2013; pp. 4011–4015. [CrossRef]

91. Saha, S.; Majumdar, A. Data centre temperature monitoring with ESP8266 based Wireless Sensor Network and cloud based dashboard with real time alert system. In Proceedings of the 2017 Devices for Integrated Circuit (DevIC), Kalyani, India, 23–24 March 2017; pp. 307–310. [CrossRef]

92. Skoplaki, E.; Palyvos, J. On the temperature dependence of photovoltaic module electrical performance: A review of efficiency/power correlations. *Sol. Energy* **2009**, *83*, 614–624. [CrossRef]

93. Dubey, S.; Sarvaiya, J.N.; Seshadri, B. Temperature Dependent Photovoltaic (PV) Efficiency and Its Effect on PV Production in the World—A Review. *Energy Procedia* **2013**, *33*, 311–321. [CrossRef]

94. Skoplaki, E.; Boudouvis, A.; Palyvos, J. A simple correlation for the operating temperature of photovoltaic modules of arbitrary mounting. *Sol. Energy Mater. Sol. Cells* **2008**, *92*, 1393–1402. [CrossRef]

95. Emery, K.; Smith, R. *Monitoring System Performance*; Technical Report; National Renewable Energy Laboratory (NREL): Golden, CO, USA, 2011.

96. Maxim Integrated Products, Inc. DS18S20 DATASHEET. Available online: https://datasheets.maximintegrated.com/en/ds/DS18S20.pdf (accessed on 28 November 2018).

97. Electronic, H. RFM Datasheet. Available online: http://www.hoperf.com/upload/rf/RFM95969798W.pdf (accessed on 28 November 2018).

98. Van den Bossche, A.; Dalcé, R.; Val, T. OpenWiNo: An open hardware and software framework for fast-prototyping in the IoT. In Proceedings of the 2016 23rd International Conference on Telecommunications (ICT), Thessaloniki, Greece, 16–18 May 2016; pp. 1–6. [CrossRef]

99. Wendt, T.; Volk, F.; Mackensen, E. A benchmark survey of long range (LoRaTM) spread-spectrum-communication at 2.45 GHz for safety applications. In Proceedings of the 2015 IEEE 16th Annual Wireless and Microwave Technology Conference (WAMICON), Cocoa Beach, FL, USA, 13–15 April 2015; pp. 1–4. [CrossRef]

100. IEC 61724, Photovoltaic System Performance Monitoring-Guidelines for Measurement, Data Exchange, and Analysis. 2017. Available online: https://ci.nii.ac.jp/naid/10019072640/ (accessed on 28 November 2018).

101. IBM. IBM LMIC Framework. Arduino Port of the LMIC (LoraWAN-in-C, Formerly LoraMAC-in-C) Framework Provided by IBM. Available online: http://www.research.ibm.com/labs/zurich/ics/lrsc/lmic.html (accessed on 28 November 2018).

102. ETSI. EN300.220 Short Range Devices (SRD) Operating in the Frequency Range 25 MHz to 1000 MHz; Part 1 Technical Characteristics and Methods of Measurement. Available online: https://www.etsi.org/deliver/etsi_en/300200_300299/30022001/03.01.01_60/en_30022001v030101p.pdf (accessed on 28 November 2018).

103. Kooijman, M. LoRa(WAN) Airtime Calculator. Available online: https://docs.google.com/spreadsheets/d/1voGAtQAjC1qBmaVuP1ApNKs1ekgUjavHuVQIXyYSvNc/edit?usp=sharing (accessed on 28 November 2018).

104. Adelantado, F.; Vilajosana, X.; Tuset-Peiro, P.; Martinez, B.; Melià-Seguí, J.; Watteyne, T. Understanding the limits of LoRaWAN. *IEEE Commun. Mag.* **2017**, *55*, 48–54. [CrossRef]

105. Dragino. LoRa Wireless Products Family. Available online: http://wiki.dragino.com/index.php?title=Main_Page#LoRa_Wireless_Products_Family (accessed on 28 November 2018).

106. Foundation, R.P. Raspbian. Available online: https://www.raspberrypi.org/downloads/raspbian/ (accessed on 28 November 2018).

107. Telkamp, T. Single Channel LoRaWAN Gateway. Available online: https://github.com/tftelkamp/single_chan_pkt_fwd (accessed on 28 November 2018).

108. Network, T.T. Learn How to Grow the Network and Connect All Things. Available online: https://www.thethingsnetwork.org/docs/ (accessed on 28 November 2018).

109. Product Brief—Sentrius RG1xx Series Gateway. Available online: https://www.mouser.es/datasheet/2/223/Product%20Brief%20-%20Sentrius%20RG1xx%20Series%20Gateway-1113271.pdf (accessed on 28 November 2018).

110. Meijers, J. TTN Mapper. Available online: http://ttnmapper.org (accessed on 28 November 2018).

111. Sanchez-Iborra, R.; Sanchez-Gomez, J.; Ballesta-Vinas, J.; Skarmeta, A. Performance Evaluation of LoRa Considering Scenario Conditions. *Sensors* **2018**, *19*, 772. [CrossRef] [PubMed]

112. Augustin, A.; Yi, J.; Clausen, T.; Townsley, W. A Study of LoRa: Long Range & Low Power Networks for the Internet of Things. *Sensors* **2016**, *16*, 1466. [CrossRef]

113. SemTech. SemTech Manual Online. Available online: http://semtech.force.com/lora/LC_Answers_Questions?id=90644000000PmGvAAK (accessed on 28 November 2018).

energies

MDPI

Article

A Novel Direct Load Control Testbed for Smart Appliances

Joaquín Garrido-Zafra [1], Antonio Moreno-Munoz [1,*], Aurora Gil-de-Castro [1], Emilio J. Palacios-Garcia [2], Carlos D. Moreno-Moreno [1] and Tomás Morales-Leal [3]

[1] Electronics and Computer Engineering Department, University of Córdoba, Córdoba 14071, Spain
[2] Department of Energy Technology, Aalborg University, Aalborg 9220, Denmark
[3] Electrical Engineering Department, University of Córdoba, Córdoba 14071, Spain
* Correspondence: amoreno@uco.es; Tel.: +34-957-218373

Received: 23 July 2019; Accepted: 28 August 2019; Published: 29 August 2019

Abstract: The effort to continuously improve and innovate smart appliances (SA) energy management requires an experimental research and development environment which integrates widely differing tools and resources seamlessly. To this end, this paper proposes a novel Direct Load Control (DLC) testbed, aiming to conveniently support the research community, as well as analyzing and comparing their designs in a laboratory environment. Based on the LabVIEW computing platform, this original testbed enables access to knowledge of major components such as online weather forecasting information, distributed energy resources (e.g., energy storage, solar photovoltaic), dynamic electricity tariff from utilities and demand response (DR) providers together with different mathematical optimization features given by General Algebraic Modelling System (GAMS). This intercommunication is possible thanks to the different applications programming interfaces (API) incorporated into the system and to intermediate agents specially developed for this case. Different basic case studies have been presented to envision the possibilities of this system in the future and more complex scenarios, to actively support the DLC strategies. These measures will offer enough flexibility to minimize the impact on user comfort combined with support for multiple DR programs. Thus, given the successful results, this platform can lead to a solution towards more efficient use of energy in the residential environment.

Keywords: demand response; direct load control; home energy management system; mixed-integer linear programming

1. Introduction

Much has been written about the new role consumers can play in future smart grid (SG). Driven by the massive integration of renewable energy resources, the SG is evolving swiftly, causing changes in how electricity is produced, managed, marketed, and consumed. If for a while, the SG paradigm meant merely accepting a bi-directional flow of electricity and information, it must continue to evolve to adapt to the current demands of the digital consumer. In the years to come, the computational exploitation of the enormous amounts of information provided by the Internet of Things (IoT) sensors, incorporated at all layers of the SG, will become the main engine of its evolution towards the digital energy network, focused on customer service. This is what has been called "data-driven energy" [1]. A large amount of energy data will support collective decision making, opening the way to more responsive utilities and more engaged consumers. This will undoubtedly impact the evolution of household appliances. In fact, SAs are already showing their potential for data-driven energy [2].

The growing use of energy by domestic appliances shows no signs of slowing, reaching 2900 TWh in 2017. The use of electricity by these loads continues to grow by almost 2% per year, a steady trend since 2010 [3]. Although the electricity demand for major appliances has slightly decreased since 2007, mainly due to improvements in their energy efficiency, the rapid proliferation of small appliances and brown goods has absorbed these savings. The energy consumption due to these small loads has grown twice as fast as that of large appliances in the last decade. In addition, only one-third of domestic appliances consumption is under regulatory protection, particularly in emerging markets. This may become even more relevant in the near future as the demand for electricity in buildings increases due to the impact of the charging infrastructure for electric vehicles. While it is true that there is a need to increase the rigor of existing policies by extending regulatory coverage to a broader range of devices, on the other hand, user awareness may be the key factor. However, to achieve this, consumers should be rewarded to some extent when changing their behavior. The availability of information and communication technologies (ICT) on SG can be decisive in meeting this commitment through the widespread adoption of DR strategies.

In other areas, such as power electronics, it is common to find a complete chain of modeling, development, testing, optimization, virtual validation, and rapid prototyping commercial tools that integrate seamlessly into a convenient testing and development environment such as these tools of Typhoon (Typhon, Somerville, USA) [4] and dSPACE (dSPACE, Paderborn, Germany) [5]. It is possible to find testbed proposals for different applications in SG, like our previous one [6]. In the newly released paper [7], a distributed framework for real-time management and co-simulation of DR in SG is presented. This solution provides a near real-time co-simulation platform to validate new DR-policies exploiting IoT approach performing software-in-the-loop. In the recent papers, authors propose an interesting testbed for distributed DR based on a microgrid (MG) modeled on the PSIM software (Powersim, Rockville, USA) to provide frequency regulation [8] and control over other grid parameters in general [9]. In the model, the nodes of virtual IoT devices are created according to the collective characteristics of their real twins, connected to the system. Network conditions can be reproduced when testing new DR algorithms to provide, e.g., frequency regulation reserve services.

Similarly, in order to support the field of DLC research in this emerging application area of SA, it is necessary to provide new testbeds for lab experimentation. Therefore, the main contribution of this work is the development of a research test bench flexible enough to incorporate different tools of different origins such as weather forecasting APIs, DR providers from the utility and mathematical optimization features built on the basis of the LabVIEW systems (2015, National Instruments, Austin, USA) design platform and development environment for a visual programming language. It can benefit from user-friendly and intuitive software as well as hardware such as powerful real-time processors, user-programmable field-programmable gate array (FPGA), and full I/O interfaces. However, although it also offers libraries of dedicated functions, it has been necessary to specifically develop a sophisticated software (that did not exist) that supports the seamless link between the tools, since their individual parts are precisely aligned with each other. In this sense, the proposed testbed is a novelty since most of the papers available in the literature are focused on the development of complex mathematical models without considering the integration of these tools that are so important to implement a realistic platform and thus emulate scenarios and test cases as real as possible. Furthermore, this work is a step forward from previous research, as it includes several tools that have never been integrated before.

The organization of the paper as follows. Sections 2 and 3 presents the background of the research. Then, Section 4 describes the experimental platform and also examines the control and optimization strategies, considering practical limitations and safety constraints in detail. In Section 5, the case study is discussed. Finally, the conclusions and future work are reported in Section 6.

2. Home Energy Management Systems (HEMS) State of the Art

The combination of the SG paradigm with IoT technologies and the will of consumers to actively participate in their energy control has enhanced the HEMS concept. These are systems capable of monitoring home consumption at different levels and implementing automation or control mechanisms.

They have evolved at an unstoppable pace in the last years. By 2013, most systems only offered home monitoring, either local or remote and rarely some manual control over switches or dimmable loads [10].Currently, on the contrary, a wide variety of control systems are available ranging from the simple automatic scheduling of applications to the optimization of energy resources, through advanced algorithms that consider the state of numerous external variables such as energy prices or weather conditions. What is more, they are even able to learn from users thanks to the incorporation of artificial intelligence [11].

Because of this evolution and the large range of devices and algorithms that are being integrated into the HEMS, the number of works in the literature is extensive and unapproachable for a paper whose purpose is not that. However, for example, the authors in [12] define a classification according to the level of complexity of these systems. This will help to situate the present work and the challenges addressed. The levels from the lowest to the highest complexity are Monitoring, Logging, Alarm, Energy Management, and DR.

Nowadays, the first three levels can be regarded as a prerequisite. Every HEMS must carry out home monitoring at different aggregation levels. The basic level is the total household consumption, generally measured by technologies such as Smart Meters, widely deployed across Europe [13]. Nevertheless, the energy footprint of individual elements can be recorded by means of load submetering or non-intrusive load monitoring (NILM) algorithms, which use machine learning to distinguish individual appliances from the total consumption [14].

The capture of measures can be performed with different granularity and be stored in different supports. In this way, all or part of the data is stored in the cloud, from where it is possible to obtain descriptors or apply machine learning algorithms. This also allows for the possibility of generating alarms at different levels, so fast events that require immediate attention can be generated and then processed in the so-called Edge, while more complex alarm mechanisms can be implemented in higher layers after preprocessing and analysis of historical data.

The aforementioned elements are essential for the creation of reliable controls at the next levels named: energy management and DR. The first focuses on the control of a combination of distributed resources to guarantee a continuous power supply, whereas the second goes a step further and manages the individual consumer appliances.

Among the recent publications, the most used optimization techniques are mixed-integer linear-programming [15], and variation of those [16], as well as population-based algorithms [17]. It is also common to find works that propose multi-objective algorithms to reach a trade-off between the energy savings that can be achieved and the benefits from possible incentives [18].

Nevertheless, as is evident from the most recent publications, the use of Internet technologies as a solution to optimization problems is becoming more and more common [19], as they tackle issues such as the diversity of household appliances, the simultaneous pursuit of several objectives in parallel, and the uncertainty in predicting conditions such as occupancy levels, energy consumption or weather conditions [20].

3. Smart Appliances Overview

What is a SA? There is more than one definition, but popularly a SA is recognized because it has some degree of embedded processing and wireless connectivity. Sometimes called a Net appliance, an Internet appliance or an information appliance, it can be as simple as an application that warns you whenever your refrigerator has a door opened, or as complex as remotely controlling your oven from your smartphone or via a voice assistant. However, in the framework of the SG, the term "smart" focuses on those systems (with communications-enabled) which are able to modulate their electricity consumption in response to external signals such as price information [21], local measurements [22] or direct control commands [23]. In other words, those appliances that can support grid flexibility because they have been configured to respond to DR requests.

In a recent survey [24], 28% of people find SA very attractive, but people are more reluctant to buy them because of price concerns, so 49% of people say this is a barrier to buying. Other barriers include dynamic pricing, lack of interoperability and legal framework as reported by the European parliament [25] and the world economic forum [26]. First, the lack of a dynamic pricing model to the clear majority of customers is an obstacle. Users will not be willing to change their habits if they cannot perceive that this intelligent functionality can bring them substantial financial savings. Second, the high purchase premiums and long replacement cycles of these devices are prolonging their mass adoption. Thirdly, to enable the communication between SAs that use different protocols and standards, and to ensure interoperability, the communication interface must be supported by a data model that conforms to a harmonized reference ontology. A semantic platform called OpenFridge has recently been proposed in [27] that has been deployed and evaluated with real-life users distributed globally. But the candidate for such a reference ontology will almost certainly be the Smart Appliances REFerence ontology (SAREF) [28]. SAREF4ENER [29] is the SAREF extension to be able to fully support DR for the energy domain.

Finally, the lack of a clear legal structure around customer data limits growth in this area. This could include the appliance energy use pattern meaning when, how much and how is energy being consumed. These data could even be monetized. For example, appliance manufacturers might be willing to pay an energy supplier a fee for these data, as they can be of great value in terms of customer service, product support, as well as maintenance. In the case of aggregation [30], how this data could be shared among customers to allow, e.g., for their energy efficiency comparison. An aggregator can operate on behalf of a group of consumers, having access to data and possible remote adjustment over consumers' appliances. If the security of connected devices used in aggregation is not safeguarded, consumers could be exposed to several risks like data theft or request of appliance ransomware. Security flaws and data privacy issues are main concerns of the users, and only a few regions have well-defined rules about who can access, own, and share utility customer data.

However, the prospects for SA are bright. The global market for SA is projected to reach $38.35 billion by 2020, with a compound annual growth rate (CAGR) of 16.6% over the projected period 2015–2020. IoT-enabled devices (currently low, about 5% of white goods) are expected to grow dramatically, and the number of sensors is expected to increase six-fold by 2020. So, according to the International Energy Agency, by 2040 almost 1 billion households and 11 billion SA could participate in interconnected electricity systems.

Typically, DR policies can be classified between load-shifting strategies, which move the load from on-peak or event hours when demand and rates are the highest, to off-peak hours when rates are lower, and load-shedding strategies, which directly reduce or avoid energy use during on-peak hours altogether. Consequently, in the residential sector, the loads can be divided into non-shiftable, time-shiftable and energy "sheddable". The time-shiftable loads are the appliances whose operation can be moved from peak to off-peak times with the minimal loss of comfort for the inhabitant. This is the category of 'wet' appliances, e.g., dishwashers (DW), washing machines (WM), and tumble dryers (TD). These appliances account for a significant proportion of household energy consumption. Alternatively, non-shiftable loads, such as lighting and brown appliances, cannot delay their operation [31]. At present, there is no deployed infrastructure that allows remote activation of these appliances. However, their behavior has been deeply studied and it is now possible to understand the potential of the DR in supporting the operation of the network [32].

Among household appliances, a special category is the thermostatically-controlled loads (TCL) (e.g., electric water heaters (EWH), HVAC systems, refrigerators, and freezers) as their thermal inertia allows for flexible load patterns (both shifting and shedding) while meeting their service requirement. Therefore, compared to other SA, TCL exhibit predictable behavior from the DR point of view, and even more when aggregated in large population clusters [33]. In recent work, a stochastic model has been presented for the generation of high temporal resolution synthetic profiles of the consumption of these domestic appliances [34]. However, its potential for flexibility remains virtually unknown. [35] presents the recent projects that are facilitating the transition from research to development. In general terms, and due to their inherent characteristics, there are two types of TCL, with different operating principles. First, resistive loads (i.e., heat generation equipment) and, second, compressor-driven loads (i.e., heat pumping equipment). Although this paper is particularly dealing with resistive loads, greater demand elasticities could be achieved if the control strategy achieved were extended to the rest of the residential TCL.

4. Structure of the Smart Appliance Control Testbed

The proposed control platform is composed of four main blocks that collect data and exchange information between each other aiming to implement the abovementioned DR policies through DLC. The platform architecture is shown in Figure 1, where LabVIEW works as the core application by handling the data provided by the outer blocks. This central block also has the highest priority from the call handling point of view, that is, LabVIEW follows the classical scheme where the main application deals with the so-called subVI to allow modular designs. At the same time, this subVIs will be the interfaces with the rest of the blocks.

The block on the right side is related to the API that provides the testbed with both weather information (Ambient temperature and photovoltaic (PV) production forecast) and the price of the energy.

Finally, these DR policies must be mathematically translated into an optimization model which includes several constraints related to the people's habits, the availability of energy from different sources and the household appliances features among others. The model should also offer a certain degree of flexibility with respect to the number of invokes and formulation changes. All these reasons have contributed to opt for General Algebraic Modelling System (GAMS) as the software used to solve the proposed model. Furthermore, another component including the functions given by the GAMS API is used to integrate this software into LabVIEW using a dynamic link library (DLL). The following sections will describe these previous blocks and their interactions in more detail.

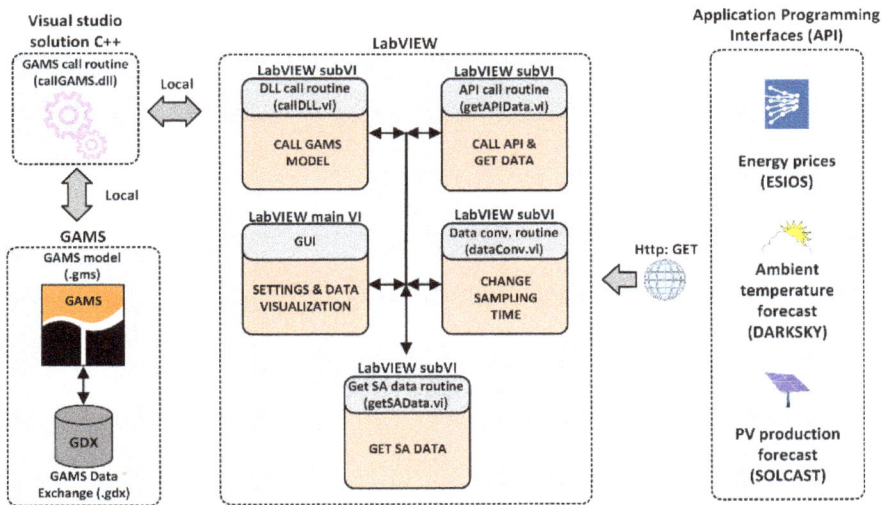

Figure 1. Smart appliances (SA) control testbed overview.

4.1. LabVIEW

LabVIEW has been the tool used to integrate and manage all blocks of the platform. Concretely, the developed LabVIEW application consists of two threads commonly known as *while loops* located in the block diagram. The first one implements the whole infrastructure necessary to parametrize and call the GAMS model and comprises the data collection from the API solcast [36], dark sky [37] and Spanish system operator information system (E-SIOS) [38], which is the information system of the Spanish electricity group *Red Eléctrica de España (REE)*, using the subVI getAPIData.vi that implements an hypertext transfer protocol (HTTP) client. This loop also entails the data standardization with respect to the sampling times by means of *dataConv.vi*, the model call through callDLL.vi as will be described in the following section and the display of the results. However, the second loop just converts the raw information of the scheduled SA (name, operation mode, and time) into a recognizable information by the model through getSAData.vi.

The graphic user interface (GUI) or front panel is shown in Figure 2 and has three main parts, namely, the SA scheduler (part A) including at the top the EWH section where the parameters that model this appliance (Minimum and maximum temperatures, tank capacity, nominal power, initial temperature, loss factor, inlet water temperature, and the hourly hot water consumption) are set up. The rest of SA under analysis in this study (WM, DW, and TD) are modeled according to their average power consumption and are scheduled at the bottom of part A where both the time and mode of operation as well as the cycle time can be selected. Two modes of operations have been evaluated: The fixed mode is used to launch the SA at a fixed time while the variable mode enables a certain degree of flexibility since the SA is scheduled over a time interval. As a result, the platform is forced to decide the start time within this interval once the model is solved. This part also includes information (Name, operation mode and time) about the scheduled SA.

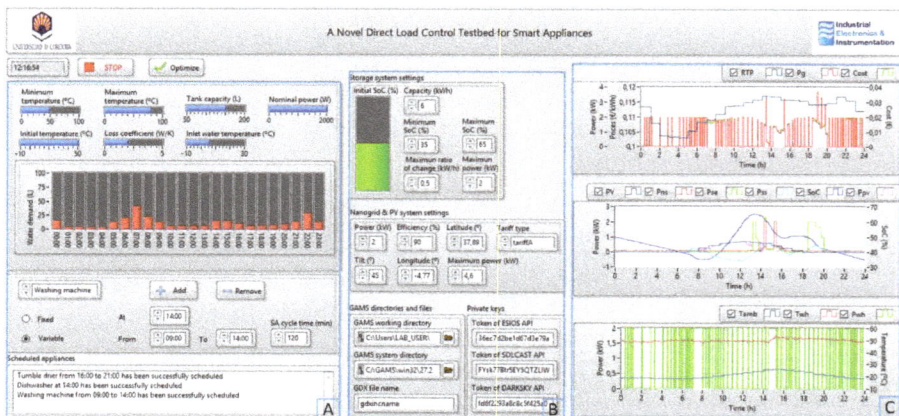

Figure 2. LabVIEW graphic user interface (GUI).

At the top of the part B are the parameters that model the energy storage system (ESS) such as the initial state of charge (SoC), capacity, minimum, and maximum SoC allowed as well as maximum power flow and a maximum ratio of change. At the middle, some features of the nanogrid under study can be found: Geographical coordinates (Longitude and latitude), tilt angle, power and efficiency in the case of the PV system or maximum power and tariff type with respect to the grid connection. The last section in part B includes the local directories required by GAMS, on the left half, and the personal keys that API administrator provides to establish a secure connection, on the right half.

Finally, part C shows the results of the optimization process divided into three graphs. From top to bottom, the first graph plots the hourly price of the energy according to the selected tariff, the optimal power consumption from the grid and thus the cost once these two previous ones are known. The second graph shows profiles such as the SoC and power taken from the batteries, the PV production and the amount of such power that would be injected into the nanogrid, the power usage of the SA and the consumption to be considered non-shiftable. The last graph describes the whole state of the EWH depicting its power consumption as well as the water and ambient temperatures.

4.2. Linking GAMS and LabVIEW

This section describes the communication between GAMS and LabVIEW. Some papers show the integration of GAMS into Matlab [39] or other software like LabVIEW through Matlab as an intermediary interface [40]. In this sense, the novelty of this work is the direct coupling of both tools without using any intermediate software. On the one hand, the inner communication between the GAMS model and its GAMS Data Exchange (GDX) file has been included under the subsection GAMS as appears in Figure 1. This file is often used to store the parameters with which the model is called, as well as the model results, however, such interaction does not take place directly but will have to be handled by means of the appropriated classes and methods that the GAMS object-oriented API [41] provides resulting in the seamless integration of GAMS into any application such as LabVIEW in this case. This architecture employs the C++ API in a DLL format which is the interface that makes the linkage possible. The flowchart is shown in Figure 3. First, system and working directories have been set; the system directory refers the path where all GAMS installation files are located while the working directory refers to the path where the GAMS models and GDX files will be stored (also shown in part B of Figure 2). The second stage aims to create a database object where the parameters used in the model will be stored, but this will be carried out in the third stage. The model execution options, such as the names of the database object and the exchange file to be used are subsequently specified. The last stages are in charge of executing the model, returning the optimal values of the decision variables.

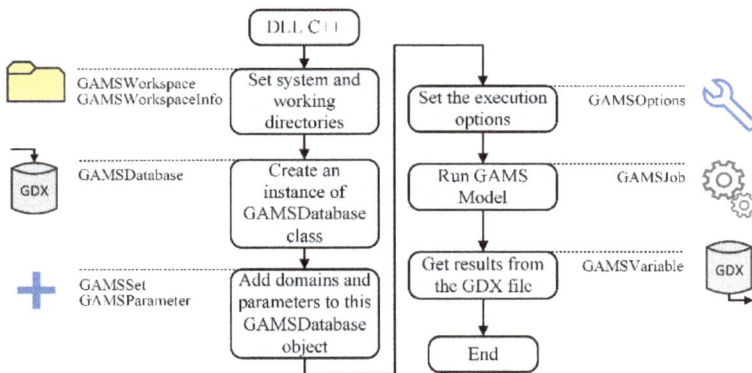

Figure 3. Dynamic-link library (DLL) flowchart for linking both environments.

4.3. An Optimization Model for Demand-Side Management

While most of the proposed models address this issue as a task scheduling problem using heuristics algorithm to make decisions on shifting, shedding or even disconnecting the load, this paper proposes a novel mixed-integer linear programming (MILP) model that uses the price-based DR programs to optimize the power consumption using the potential flexibility that TCL provides to the demand. The proposed model involves a smart home with its own ESS, distributed energy resources (DER) based on PV panels as well as a scenario with SA managed through the DLC strategy.

In terms of mathematical formulation, Equation (1) refers to the objective function f (in €) where $P_g(t)$ (in kW, as all the powers henceforth) and $Pr(t)$ (in €/kWh) are the power consumption from the grid and the price of the energy respectively at time slot $t \in [1, 2, \ldots T]$. The rest of the equations are constraints related to the power balance, the user preferences and the energy availability from the sources. Equation (2) denotes the global power balance at each time slot t with $P_g(t)$, the power taken from the PV panels $P_{pv}(t)$ as well as the power given by the energy storage system $P_{ess}(t)$ on the generation side. On the demand side is the non-shiftable power $P_{ns}(t)$, which involves the stand-by consumption coming from non-dimmable devices, such as lighting, low power DC adapters used to supply small devices and also the consumption from the SA scheduled in fixed mode, $P_s(t)$ is the power consumption of the common-use SA such as the WM, DW and TD scheduled in variable mode and the consumption of the EWH expressed as the product of its nominal power P_{wh} and the binary variable $x_{wh}(t)$ that indicates its state each time slot.

Equations (3)–(5) set the limits for $P_g(t)$, $P_{pv}(t)$ and $P_{ess}(t)$ respectively, where P_g^{max} denotes the maximum power that can be taken from the grid and P_{ess}^{max} the maximum power that can be injected into or extracted from the ESS. Moreover, the ratio of change of this variable has also been constrained in Equation (8) through the parameter dP_{ess}^{max} (in kW/h) in order to ensure a lifetime of the batteries as long as possible. Finally, $f_{pv}\left(P_{pv}^{pk}, \eta_{pv}, \alpha, \lambda, \phi, t\right)$ refers to the function that implements solcast to provide the PV production each time slot t and thus has been taken as the maximum power available to be injected into the system from the PV panels. Parameters such as the installed peak power P_{pv}^{pk}, the efficiency η_{pv}, the tilt angle α or the location, through the latitude λ and longitude ϕ will be required by this API in each HTTP request.

Equations (6) and (7) describe the dynamic of the ESS by means of a simple kWh counter to compute the current state of charge $SoC(t)$ (in %) based on the previous one $SoC(t-1)$ and $P_{ess}(t)$ and setting the $SoC(t)$ limits between SoC_{min} and SoC_{max} not to allow deep charges and discharges which is also a condition to ensure a long lifetime of the system.

The SA scheduling process using the variable mode is modeled by Equations (9) and (10) and has been conceived as a decision-maker who chooses the optimal SA operation from among the possible ones that could be generated between the selected start and end times, by shifting the original consumption one-time slot. In this context, let us define j as the index that refers to each SA to be scheduled and k_j the index associated with each shifted consumption that may be generated for each j, being N_j the number of SA and N_k^j the number of possible consumption profiles. This family of shifted consumptions builds each matrix $\Gamma_j(k_j, t)$ which has as many rows as possible scenarios and as many columns as considered time slots T. Moreover, for the decision-making process, all the shifted consumptions have been associated with a binary variable $x_j(k_j)$ and thus, the optimal scenario will be indicated once the model is solved by means of the state of these decision variables. Finally, to ensure just one shifted consumption operates, Equation (9) forces the sum so that just one binary variable is equal to 1.

The EWH has been considered as a special SA due to its thermal inertia and therefore has its own power balance equation as it is apparent from (11). From left to right, this balance involves the energy stored inside the EWH tank characterized by the current and previous average water temperature $T_{wh}(t)$ and $T_{wh}(t-1)$ (in °C, as the rest of temperatures hereafter), the tank capacity C_{wh} (in m^3) and the parameters that model essential features of the supply water like its density ρ (in kg/m^3) and its specific heat C_p (in kJ/kg·°C). The following terms are the thermal losses taking place in the tank walls given by the loss factor g_{wh} (in kW/°C) and the ambient temperature profile $T_{amb}(t)$ besides the energy provided by the water entering the tank as a consequence of the usage events and defined by means of the hot water consumption $D_{wh}(t)$ (in m^3/s) and the temperature of this water, T_{inlet}. Finally, the discrete energy due to the heater element can be found. Once the EWH dynamic has been well-defined, the model for this appliance is fully completed with Equation (12) where the upper and lower limit of $T_{wh}(t)$ are constrained according to the normal operation temperatures T_{wh}^{min} and T_{wh}^{max}.

$$Min\ f = \frac{24}{T} \sum_{t=1}^{T} P_g(t)Pr(t) \tag{1}$$

S.t:

$$P_g(t) + P_{pv}(t) + P_{ess}(t) = P_{ns}(t) + P_s(t) + x_{wh}(t)P_{nwh} \tag{2}$$

$$0 \leq P_g(t) \leq P_g^{max} \tag{3}$$

$$0 \le P_{pv}(t) \le f_{pv}\left(P_{pv}^{pk}, \eta_{pv}, \lambda, \phi, t\right) \tag{4}$$

$$-P_{ess}^{max} \le P_{ess}(t) \le P_{ess}^{max} \tag{5}$$

$$SoC(t) = SoC(t-1) - 100\left(\frac{24}{T}\right)\frac{P_{ess}(t)}{C_{ess}} \tag{6}$$

$$SoC^{min} \le SoC(t) \le SoC^{max} \tag{7}$$

$$-dP_{ess}^{max} \le \frac{P_{ess}(t) - P_{ess}(t-1)}{\frac{24}{T}} \le dP_{ess}^{max} \tag{8}$$

$$P_s(t) = \sum_{j=1}^{N_j} \sum_{k_j=1}^{N_k^j} x_j(k_j)\Gamma_j(k_j, t) \tag{9}$$

$$\sum_{k_j=1}^{N_k^j} x_j(k_j) = 1 \tag{10}$$

$$C_{wh}\rho C_p \frac{T_{wh}(t) - T_{wh}(t-1)}{\frac{86400}{T}} =$$
$$g_{wh}[T_{amb}(t) - T_{wh}(t-1)] + D_{wh}(t)\rho C_p[T_{inlet} - T_{wh}(t-1)] + x_{wh}(t)P_{nwh} \tag{11}$$

$$T_{wh}^{min} \le T_{wh}(t) \le T_{wh}^{max} \tag{12}$$

5. Case Study

This section aims to evidence the effectiveness of the above-described DLC platform by testing it under cases which consist of minimizing the cost of the energy imported from the grid over a 24-h time horizon, as was stated in the previous section, with a time resolution of 5 min, so that, $T = 288$. Concretely, two case studies based on the available electricity tariffs in the Spanish market are considered. One case is based on time discrimination in two periods (off-peak and peak) also known as tariff DHA, and another is a case using the default tariff or tariff A (without time discrimination).

Both cases use the SA consumption models shown in Figure 4 and based on 120 min working cycle divided into 8 slots of 15 min provided by [42].

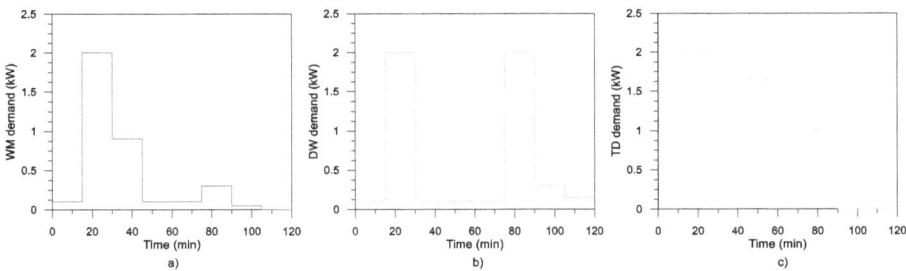

Figure 4. Smart appliances (SA) models employed in the optimization: (**a**) Washing machine demand, (**b**) dishwasher demand, and (**c**) tumble drier demand.

Additionally, to give the case study a more realistic approach, the component of the non-shiftable power that represents the standby consumption was obtained by acquiring the active power in one of the laboratory circuits for a 24-h workday.

Under this framework, a typical dwelling including a small scale ESS and PV installation has been chosen as the topology of this case study. More in detail, the PV system is modeled by a nominal power of 2 kW, an efficiency of 90% and mounted with a tilt angle of 30°. With respect to the location, southern

Spain has been considered for both cases, concretely at 37.88° and −4.79° of latitude and longitude respectively. On the other hand, the ESS has a capacity of 6 kWh, a maximum charge/discharge power of 2 kW with a ratio of change limited to 0.5 kW/h and where the SoC can fluctuate in the range 35–65%, the initial SoC was fixed to 50%. The EWH considered is the type which can be found in the residential environment, vertically mounted and cylindrical, with a capacity of 0.1 m³ as well as a nominal power of 2 kW. Its loss factor has been set to 2·10⁻³ kW/°C and the inlet water temperature to 21 °C [43], while the water temperature inside the tank has been constrained in the range 60–85 °C with an initial condition of 65 °C. In addition, an example of hot water consumption considering the water drawn from the EWH tank due to household use such as hand washing, showering, and dishwashing among others and based on [44] has been used. Finally, the capacity of the main grid has been fixed to 4.6 kW since it is a common value in Spain. Table 1 summarizes the main parameters of the model as well as its values.

Table 1. Main parameters of the model.

Subsystem	Parameter	Value	Subsystem	Parameter	Value
Main grid	P_g^{max}	4.6 kW		C_{wh}	0.1 m³
ESS	P_{ess}^{max}	2 kW	EWH	g_{wh}	2·10⁻³ kW/°C
	C_{ess}	6 kWh		T_{inlet}	21 °C
	dP_{ess}^{max}	0.5 kW/h		T_{wh}^{min}	60 °C
	SoC^{min}	35%		T_{wh}^{max}	85 °C
	SoC^{max}	65%		$T_{wh}(0)$	65 °C
	$SoC(0)$	50%		P_{nwh}	2 kW
PV	P_{pv}^{pk}	2 kW		C_p	4.18 kJ/kg·°C
	η_{pv}	90%		ρ	988 kg/m³
	α	30°			
	λ	37.88°			
	ϕ	−4.79°			

Figure 5 introduces the first case in which tariff A with both scheduling mode (variable and fixed) have been used, depicting a 24-h horizon. Particularly, in this case, the scheduling configuration for the SA has been set as follows: Washing machine scheduled from 09:00 to 14:00, tumble dryer scheduled from 16:00 to 21:00 and dishwasher fixed at 14:00.

Note from Figure 5a the result of the scheduling process and the times at which the SA start their operation cycles. As it is apparent from $P_{sa}(t)$, which is the decoupled consumption of all the SA scheduled in either fixed or variable mode, the washing machine starts almost at midday (at 11:45), around the peak of the prices although a large amount of this demand is covered by the PV system. The dishwasher at 14:00 (as was stated) and tumble drier is shifted until 18:00 where the second valley of the price can be found. This behavior shows a clear strategy of searching for the lowest price or the highest PV production to launch these SA. In view of the results, all the initial constraints related to the scheduling period are clearly satisfied.

The non-shiftable consumption is denoted by the red line of the same figure including the fixed consumption of the dishwasher at 14:00 and the experimentally measured example in which the period of highest activity falls in the range 09:00–18:00 according to the laboratory timetables. The green line shows the power injected into the system from the PV panels, which represents 9.78 kWh, and has not the same value that the PV production shown in orange (10.65 kWh) and provided by solcast. In this case, the system does not use all the energy to achieve the most economical way, however, the amount of this one taken from the main grid is greater than if the PV energy were fully employed.

Figure 5b depicts the EWH behavior using the above-mentioned hot water demand (expressed in L/h instead of m³/s for easier comprehension) and the hourly temperature profile provided by dark sky (see purple and blue lines respectively). The EWH consumption shown in orange evolves in the range 0–2 kW due to it's on/off operation. Before 12:00, the water temperature is more or less constant and the power consumption behaves in agreement to the water consumption so that a water demand variation causes a proportional energy consumption, which means this energy is mainly used to warm the inlet water. In fact, the highest energy consumption in this interval takes place at the peak of water demand. On the contrary, at midday, the water consumption is not significative and thus, this energy is intended to increase the water temperature inside the tank from 60 °C to 83 °C, considering multiple favorable conditions such as the greater availability of energy coming from the PV system, the high ambient temperature as well as the amount of charge already stored in the ESS. This temperature increment enables to face the future water drawn acts, which is a desirable strategy in response to DR events as it is the presence of high market prices in this interval. Later, the temperature slowly falls up to 60 °C at 20:00 due to the water consumption and remains constant the rest of the day.

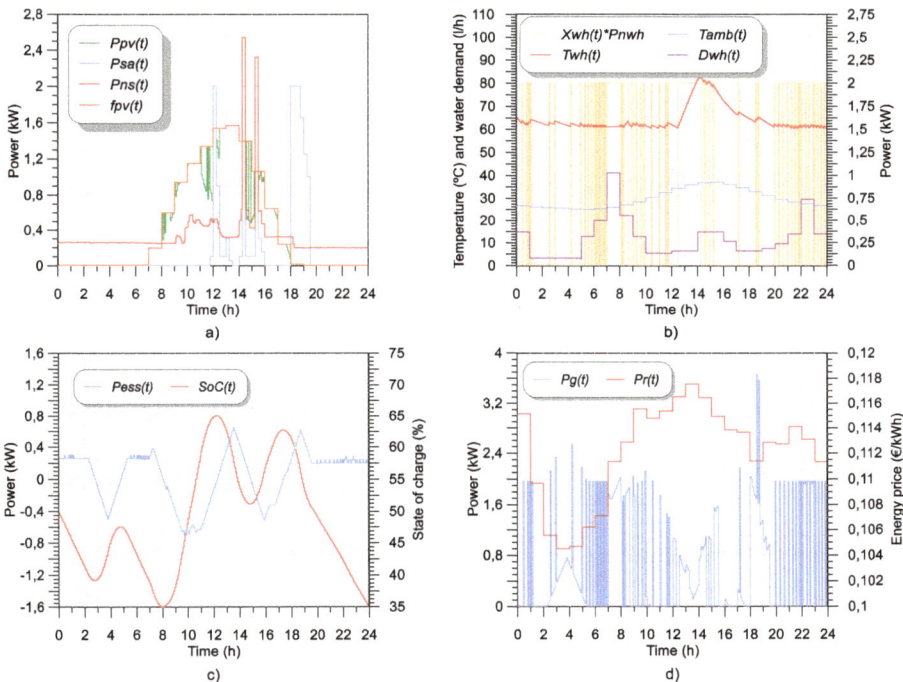

Figure 5. Optimization results using tariff A: (**a**) Power injected by the photovoltaic (PV) system, SA consumption, non-shiftable consumption, and PV production, (**b**) EWH performance: Consumption, ambient and water temperature as well as hot water demand, (**c**) energy storage system (ESS) performance: Power and state of charge (SoC), and (**d**) total consumption from the utility and energy prices.

The ESS shows a clear policy based on the energy price (red line in Figure 5d) and PV production. The initial SoC was set to 50% and quickly decreases to supply the non-shiftable power until 02:00 reaching almost 39% in a high-priced environment. Afterward, the off-peak of the prices can be found and $P_{ess}(t)$ go up as fast as possible (due to the slope of $P_{ess}(t)$ matches to dP_{ess}^{max}) to retrieve some charge previously lost, which is equivalent to shift the amount of energy that belongs to $P_{ns}(t)$, from the beginning of the day to the off-peak interval. During this interval $P_{ns}(t)$ is supplied by means of the main grid. At 04:30 the SoC drops again to repeat the same process with $P_{ns}(t)$ and reaches the minimum value allowed (35%) at 08:00. Once here, the PV system begins to inject power that goes directly to the ESS resulting in a charging process that carries the SoC from 35% to 65% to address the

SA consumption with the help of $P_{pv}(t)$ during the most expensive interval (12:00–15:00). The rest of time follows the same principle as explained above: Charge process in presence of the second off-peak of the price (15:00–17:30) and subsequent discharge to supply both the tumble drier and the non-shiftable power (17:30–00:00). The total energy exported and imported by the system was 3.89 and 3.00 kWh respectively. Another important detail is the effect of dP_{ess}^{max} over $P_g(t)$: Previous tests were done with a more relaxed value prove that a larger amount of the EWH energy can be absorbed by ESS as this would enable better tracking of demands with higher ratios of change. Finally, in Figure 5d the hourly prices and the main grid consumption can be found. The foregoing description of this case is also reflected in $P_g(t)$ and makes it possible the main objective of avoiding and capitalizing the peak and off-peak of $Pr(t)$ respectively. The daily price for this case was 1.80 € with a total demand of energy that almost achieves 16.10 kWh.

For the following case, the configuration for the SA has been set as follows: Washing machine fixed at 10:00, tumble dryer scheduled from 12:00 to 20:00 and dishwasher scheduled from 14:00 to 19:00.

Figure 6a (blue line) shows how the model has decided to launch the dishwasher and tumble drier at the lower limit of the scheduling period which allows the system benefits from the PV production (depicted in orange and kept constant from the previous case) and the ESS that also supplies part of this consumption, especially after 13:00, where the prices are much higher than in the previous half. The washing machine operates at 10:00 as expected. The PV production is not fully intended to be injected into the system (just 9.96 of 10.65 kWh) as is evidenced by $P_{pv}(t)$, in green, and which also took place in the case above. With respect to the non-shiftable demand, the previous part corresponding to the stand-by consumption has been used, including the demand of the washing machine at 10:00.

Both the ESS and EWH have similar behaviors with respect to the previous case but with some exceptions. Figure 6b shows the performance of the EWH under the same assumptions as of the first case (water demand, ambient temperature, water temperature limits, and initial conditions) although the temperature increment begins one hour earlier and is more progressive. Furthermore, the temperature rises at one of the peaks of the water demand while the water was warmed up before this maximum in the first case. The ESS also performs similar, which evidences the PV production has a higher weight in its behavior than the energy price. Moreover, with respect to $Pr(t)$, it is more important the shape of the function, concretely the maxima and minima location, than the absolute values. The energy exported and imported in this case reaches 3.45 and 2.55 kWh. Finally, Figure 6d introduces the prices, that splits the day in two well-defined half, and the consumption from the main grid where the most consumption is located in the cheapest region as desired and entails an amount of 15.75 kWh (11.8 kWh from 23:00 to 13:00 compared to 3.95 kWh the rest of time). The daily price was 1.30 €.

Once these previous cases have been presented, Table 2 summarizes the results. Obviously, case 2 achieves a better performance with respect to the objective function and thus, tariff DHA enables to more efficient utilization of elements such as DER and ESS in presence of thermal loads that contribute to the flexibility of the system as in this case the EWH.

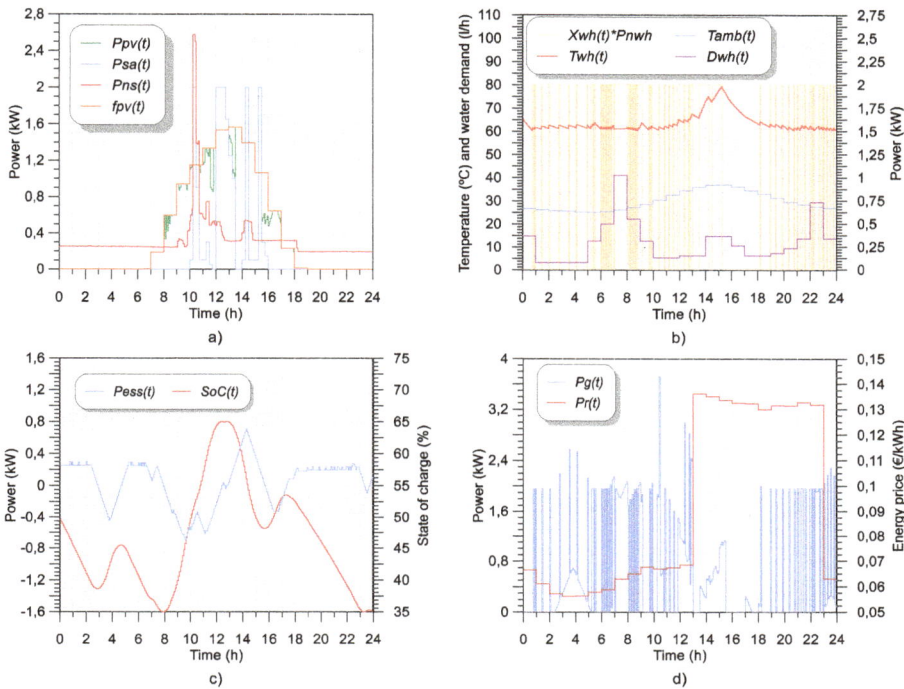

Figure 6. Optimization results using tariff DHA: (**a**) Power injected by the PV system, SA consumption, non-shiftable consumption, and PV production, (**b**) EWH performance: Consumption, ambient and water temperature as well as hot water demand, (**c**) ESS performance: Power and SoC and (**d**) total consumption from the utility and energy prices.

Table 2. Result summary.

	ESS		PV		Main Grid	
Case	Imported Energy (kWh)	Exported Energy (kWh)	Injected Energy (kWh)	Energy Production (kWh)	Energy Imported (kWh)	Objective Function: Price (€)
Case 1: Tariff A	3.00	3.89	9.78	10.65	16.10	1.80
Case 2: Tariff DHA	2.55	3.45	9.96	10.65	15.75	1.30

6. Conclusions and Future Work

In the current context of increasing energy use in the residential environment, where most consumption comes from the SA use, the employment of DR policies is essential to deal with this type of loads through a DLC paradigm with the goal of reaching higher efficient management of the energy resources. This paper has proposed an original architecture that supports research and development, and integrates tools that are very diverse and complementary aiming to develop a platform that brings together the best features of all of them, such as the high mathematical performance of GAMS, the accuracy of the weather forecasting applications as well as the flexibility of LabVIEW as the linking tool. Later both cases studies have been carried out to prove the high capabilities of the testbed with successful results, placing the adopted solution as an attractive alternative towards a higher energy performance dwelling ambient.

Finally, as future work, the authors leave the real-time control of the DER, ESS, and loads in the primary and secondary control of a real MG. In this context, the developed platform would perform as a day-ahead demand scheduler in the tertiary control although additional communication channels would need to be deployed to enable the interface with the lower hierarchical level. Furthermore, the mathematical model written in GAMS and thus the developed DLL would also have to be adapted to the MG needs, however, due to the reconfigurable nature of the system, this would not take more than a few minutes. Hence, this platform could be migrated to be used in a real microgrid expecting the same performance, but these considerations must be considered.

Author Contributions: C.D.M.-M and E.J.P.-G made a comprehensive review of the existing literature. A.M-M. and J.G.-Z. address the conception, research, and design of the work presented here. J.G.-Z. and T.M.L. were involved in the development and integration of the system's components. A.G.C. guided the whole work, edited the language, and provided their comments on the manuscript. All authors contributed to writing and reviewing the paper.

Funding: This research was supported by the Spanish Ministry of Economy and Competitiveness under Project TEC2016-77632-C3-2-R. The IMPROVEMENT project (Interreg SUDOE SOE3/P3/E0901) is acknowledged for partially funding this work.

Conflicts of Interest: The authors declare no conflict of interest.

References

1. Amasyali, K.; El-Gohary, N.M. A review of data-driven building energy consumption prediction studies. *Renew. Sustain. Energy Rev.* **2018**, *81*, 1192–1205. [CrossRef]
2. Alonso-Rosa, M.; Gil-de-Castro, A.; Medina-Gracia, R.; Moreno-Munoz, A.; Cañete-Carmona, E. Novel Internet of Things Platform for In-Building Power Quality Submetering. *Appl. Sci.* **2018**, *8*, 1320. [CrossRef]
3. Appliances and Equipment Tracking Clean Energy Progress. Available online: https://www.iea.org/tcep/buildings/appliances/ (accessed on 16 Janurary 2019).
4. dSPACE Electrical Power Systems Simulation Package. Available online: https://www.dspace.com/en/inc/home/products/sw/impsw/epc-sim-pack.cfm (accessed on 22 April 2019).
5. Typhoon HIL, Hardware in the Loop Testing Software and Hardware. Available online: https://www.typhoon-hil.com/ (accessed on 22 April 2019).
6. Moreno-Garcia, I.M.; Moreno-Munoz, A.; Pallares-Lopez, V.; Gonzalez-Redondo, M.J.; Palacios-Garcia, E.J.; Moreno-Moreno, C.D. Development and application of a smart grid test bench. *J. Clean. Prod.* **2017**, *162*, 45–60. [CrossRef]
7. Barbierato, L.; Estebsari, A.; Pons, E.; Pau, M.; Salassa, F.; Ghirardi, M.; Patti, E. A Distributed IoT Infrastructure to Test and Deploy Real-Time Demand Response in Smart Grids. *IEEE Internet Things J.* **2019**, *6*, 1136–1146. [CrossRef]
8. Thornton, M.; Motalleb, M.; Smidt, H.; Branigan, J.; Siano, P.; Ghorbani, R. Internet-of-Things Hardware-in-the-Loop Simulation Architecture for Providing Frequency Regulation With Demand Response. *IEEE Trans. Ind. Inf.* **2018**, *14*, 5020–5028. [CrossRef]
9. Thornton, M.; Motalleb, M.; Smidt, H.; Branigan, J.; Ghorbani, R. Demo abstract: Testbed for distributed demand response devices—Internet of things. *Comput. Sci. Dev.* **2018**, *33*, 277–278. [CrossRef]
10. Shafiullah, G.M.; M.t. Oo, A.; Shawkat Ali, A.B.M.; Wolfs, P. Potential challenges of integrating large-scale wind energy into the power grid—A review. *Renew. Sustain. Energy Rev.* **2013**, *20*, 306–321. [CrossRef]
11. Shareef, H.; Ahmed, M.S.; Mohamed, A.; Al Hassan, E. Review on Home Energy Management System Considering Demand Responses, Smart Technologies, and Intelligent Controllers. *IEEE Access* **2018**, *6*, 24498–24509. [CrossRef]
12. Zhou, B.; Li, W.; Chan, K.W.; Cao, Y.; Kuang, Y.; Liu, X.; Wang, X. Smart home energy management systems: Concept, configurations, and scheduling strategies. *Renew. Sustain. Energy Rev.* **2016**, *61*, 30–40. [CrossRef]
13. Zhou, S.; Brown, M.A. Smart meter deployment in Europe: A comparative case study on the impacts of national policy schemes. *J. Clean. Prod.* **2017**, *144*, 22–32. [CrossRef]
14. Abubakar, I.; Khalid, S.N.; Mustafa, M.W.; Shareef, H.; Mustapha, M. Application of load monitoring in appliances' energy management—A review. *Renew. Sustain. Energy Rev.* **2017**, *67*, 235–245. [CrossRef]

15. Babonneau, F.; Caramanis, M.; Haurie, A. A linear programming model for power distribution with demand response and variable renewable energy. *Appl. Energy* **2016**, *181*, 83–95. [CrossRef]

16. Killian, M.; Zauner, M.; Kozek, M. Comprehensive smart home energy management system using mixed-integer quadratic-programming. *Appl. Energy* **2018**, *222*, 662–672. [CrossRef]

17. Graditi, G.; Di Silvestre, M.L.; Gallea, R.; Sanseverino, E.R. Heuristic-based shiftable loads optimal management in smart micro-grids. *IEEE Trans. Ind. Inf.* **2015**, *11*, 271–280. [CrossRef]

18. Ghazvini, M.A.F.; Soares, J.; Horta, N.; Neves, R.; Castro, R.; Vale, Z. A multi-objective model for scheduling of short-term incentive-based demand response programs offered by electricity retailers. *Appl. Energy* **2015**, *151*, 102–118. [CrossRef]

19. Jindal, A.; Kumar, N.; Singh, M. Internet of energy-based demand response management scheme for smart homes and PHEVs using SVM. In *Future Generation Computer Systems*; Elsevier: Amsterdam, The Netherlands, 2018.

20. Beaudin, M.; Zareipour, H. Home energy management systems: A review of modelling and complexity. *Renew. Sustain. Energy Rev.* **2015**, *45*, 318–335. [CrossRef]

21. Department for Business, Industrial Strategy. Smart Appliances. Government Response to Consultation on Proposals Regarding Smart Appliances. The Consultation and Impact Assessmentcan be Found on the BEISsection of GOV.UK. 2018. Available online: https://assets.publishing.service.gov.uk/government/uploads/system/uploads/attachment_data/file/748115/smart-appliances-consultation-government-response.pdf (accessed on 28 August 2019).

22. Bertoldi, P.; Serrenho, T. Smart appliances and smart homes: Recent progresses in the EU. Energy Effic. Domest. Appliances Light. In Proceedings of the 9th international conference on Energy Efficiency in Domestic Appliances and Lighting, Irvine, CA, USA, 13–15 September 2017; p. 970.

23. Ectors, D.; Gerard, H.; Rivero, E.; Vanthournout, K.; Verbeeck, J.; Virag Viegand Maagøe, A.A.; Huang, B.; Viegand, J. Preparatory Study on Smart Appliances (Lot 33) Task 7-Policy and Scenario Analysis. Available online: https://eco-smartappliances.eu/sites/ecosmartappliances/files/downloads/Task_7_draft_20170914.pdf (accessed on 28 August 2019).

24. techUK. *State of the Connected Home 2018*. Available online: https://www.techuk.org/insights/news/item/13914-connected-home-device-ownership-up-but-consumers-remain-sceptical (accessed on 28 August 2019).

25. Šajn, N. Briefing Smart Appliances and the Electrical System *. 2016. Available online: http://www.europarl.europa.eu/RegData/etudes/BRIE/2016/595859/EPRS_BRI(2016)595859_EN.pdf (accessed on 28 August 2019).

26. Martin, C.; Starace, F.; Tricoire, J.P. *The Future of Electricity New Technologies Transforming the Grid Edge*; World Economic Forum: Cologny, Switzerland, 2017.

27. Fensel, A.; Tomic, D.K.; Koller, A. Contributing to appliances' energy efficiency with Internet of Things, smart data and user engagement. *Futur. Gener. Comput. Syst.* **2017**, *76*, 329–338. [CrossRef]

28. ETSI. TS 103 264-V2.1.1—SmartM2M; Smart Appliances; Reference Ontology and oneM2M Mapping. 2017. Available online: https://www.etsi.org/deliver/etsi_ts/103200_103299/103264/02.01.01_60/ts_103264v020101p.pdf (accessed on 28 August 2019).

29. ETSI. TS 103 410-1 V1.1.1 SmartM2M; Smart Appliances Extension to SAREF; Part 1: Energy Domain. 2017. Available online: https://www.etsi.org/deliver/etsi_ts/103400_103499/10341001/01.01.01_60/ts_10341001v010101p.pdf (accessed on 28 August 2019).

30. Electricity Aggregators: Starting off on the Right Foot with Consumers. Available online: https://www.beuc.eu/publications/beuc-x-2018-010_electricity_aggregators_starting_off_on_the_right_foot_with_consumers.pdf (accessed on 28 August 2019).

31. Palacios-Garcia, E.J.; Chen, A.; Santiago, I.; Bellido-Outeiriño, F.J.; Flores-Arias, J.M.; Moreno-Munoz, A. Stochastic model for lighting's electricity consumption in the residential sector. Impact of energy saving actions. *Energy Build.* **2015**, *89*, 245–259. [CrossRef]

32. Palacios-García, E.J.; Moreno-Munoz, A.; Santiago, I.; Flores-Arias, J.M.; Bellido-Outeiriño, F.J.; Moreno-Garcia, I.M. Modeling human activity in Spain for different economic sectors: The potential link between occupancy and energy usage. *J. Clean. Prod.* **2018**, *183*, 1093–1109. [CrossRef]

33. Kleidaras, A.; Kiprakis, A.E.; Thompson, J.S. Human in the loop heterogeneous modelling of thermostatically controlled loads for demand side management studies. *Energy* **2018**, *145*, 754–769. [CrossRef]

34. Palacios-Garcia, E.J.; Moreno-Munoz, A.; Santiago, I.; Flores-Arias, J.M.; Bellido-Outeirino, F.J.; Moreno-Garcia, I.M. A stochastic modelling and simulation approach to heating and cooling electricity consumption in the residential sector. *Energy* **2018**, *144*, 1080–1091. [CrossRef]

35. Kohlhepp, P.; Harb, H.; Wolisz, H.; Waczowicz, S.; Müller, D.; Hagenmeyer, V. Large-scale grid integration of residential thermal energy storages as demand-side flexibility resource: A review of international field studies. *Renew. Sustain. Energy Rev.* **2019**, *101*, 527–547. [CrossRef]

36. Detailed Data with the Simple Radiation API Tool. Available online: https://solcast.com.au/solar-data-api/api/solar-radiation-data/ (accessed on 13 March 2019).

37. Dark Sky API: Documentation Overview. Available online: https://darksky.net/dev/docs (accessed on 13 March 2019).

38. API Esios Documentation. Available online: https://api.esios.ree.es/ (accessed on 13 March 2019).

39. Wimmer, P.; Kandler, C.; Honold, J. Potential of Demand and Production Shifting in Residential Buildings by Using Home Energy Management Systems. Build Simul. 2015. Available online: http://www.ibpsa.org/proceedings/BS2015/p2821.pdf (accessed on 28 August 2019).

40. Luna, A.C.; Meng, L.; Diaz, N.L.; Graells, M.; Vasquez, J.C.; Guerrero, J.M. Online Energy Management Systems for Microgrids: Experimental Validation and Assessment Framework. *IEEE Trans. Power Electron.* **2018**, *33*, 2201–2215. [CrossRef]

41. GAMS Application Programming Interfaces. Available online: https://www.gams.com/latest/docs/API_MAIN.html (accessed on 13 March 2019).

42. Bilton, M.; Aunedi, M.; Woolf, M.; Strbac, G. Smart Appliances for Residential Demand Response (Report A10, for the Low Carbon London, LCNF Project). Imp Coll. London. 2014. Available online: https://pdfs.semanticscholar.org/e3b8/ebf700ca1317b98dca21f04bd1e3629288b7.pdf (accessed on 28 August 2019).

43. IDEA. Guía Técnica Agua Caliente Sanitaria Central. Available online: https://www.idae.es/uploads/documentos/documentos_08_Guia_tecnica_agua_caliente_sanitaria_central_906c75b2.pdf (accessed on 28 August 2019).

44. American Society of Heating, Refrigerating and Air Conditioning Engineers. *ASHRAE Handbook: HVAC Applications*; American Society of Heating, Refrigerating and Air-Conditioning Engineers: Atlanta, GA, USA, 2007.

MDPI

St. Alban-Anlage 66

4052 Basel

Switzerland

Tel. +41 61 683 77 34

Fax +41 61 302 89 18

www.mdpi.com

Energies Editorial Office

E-mail: energies@mdpi.com

www.mdpi.com/journal/energies

www.ingramcontent.com/pod-product-compliance
Lightning Source LLC
Chambersburg PA
CBHW051912210326
41597CB00033B/6120